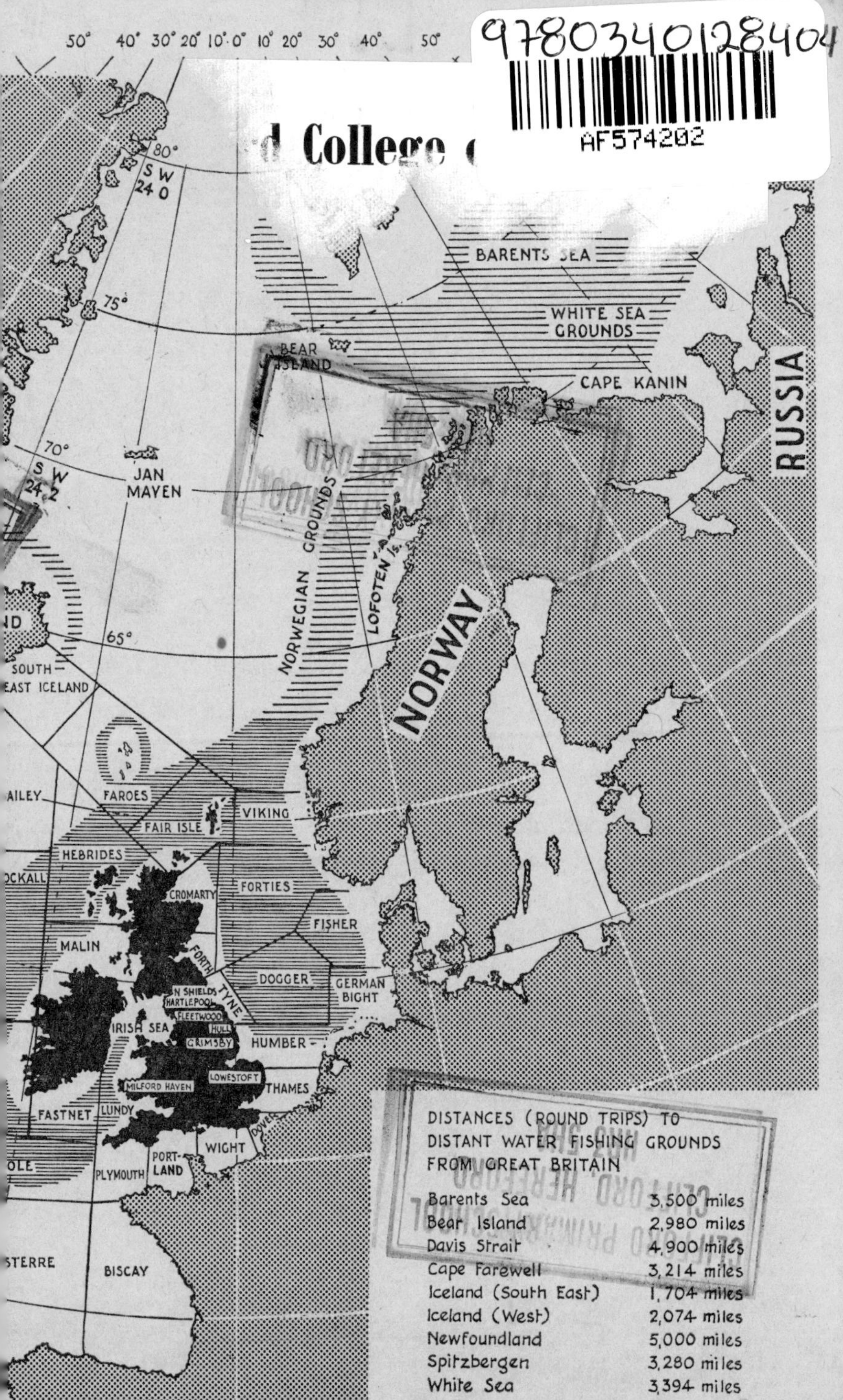

DISTANCES (ROUND TRIPS) TO DISTANT WATER FISHING GROUNDS FROM GREAT BRITAIN

Barents Sea	3,500 miles
Bear Island	2,980 miles
Davis Strait	4,900 miles
Cape Farewell	3,214 miles
Iceland (South East)	1,704 miles
Iceland (West)	2,074 miles
Newfoundland	5,000 miles
Spitzbergen	3,280 miles
White Sea	3,394 miles

EX LIBRIS
IB HAGEN
1927-1997

The Deep Sea Fishermen

By the same author:

BY WAY OF CAPE HORN

CRUISE OF THE CONRAD

GIVE ME A SHIP TO SAIL

SONS OF SINBAD

THE SET OF THE SAILS

THE QUEST OF THE SCHOONER ARGUS

CAPTAIN COOK: THE SEAMEN'S SEAMAN

THE CUTTY SARK

The Deep Sea Fishermen

Alan Villiers

HODDER AND STOUGHTON
LONDON SYDNEY AUCKLAND TORONTO

 ISBN 0 340 12840 2. *Printed in Great Britain for Hodder and Stoughton Limited, St. Paul's House, Warwick Lane, London, E.C.4., by Elliott Bros. and Yeoman Ltd., Speke, Liverpool.*

To

Sir William Smith Duthie, O.B.E.

Chairman of the

Royal National Mission to Deep Sea Fishermen

During the Vital Years

of its Greatest Progress

This Century

1954—

Contents

Illustrations

Acknowledgements

1 Alan Villiers

2 *Northern Scott*, Elgin

3 National Maritime Museum

4 Royal National Mission to Deep Sea Fishermen

Foreword

THIS is not the story of deep sea fishing or fishermen. There is a library on the subject and I am not competent to add to it. This is rather a quiet picture—a glimpse—of that immense, demanding, invaluable and sometimes dangerous industry as I have seen and experienced a little of it over several voyages. I am grateful to Mr. T. W. Boyd, C.B.E., D.S.O., once of the Royal Navy's Coastal Forces, now head of Boyd Line of Hull, for inviting me to make a White Sea voyage in his distant waters-man *Arctic Vandal*, and I thank Skipper Bernard Spicer and his men for so graciously accepting me there.

Nor is this brief work anything like an adequate picture of that lively, tremendous, and most Christian organisation, the Royal National Mission to Deep Sea Fishermen, and its cheerful, indefatigable workers. Of this I have given a glimpse too, though hurried. I am glad to have had a chance to see something of its work, and I am grateful to the Chairman, Sir William S. Duthie, Kt., O.B.E., and the Deputy Chairman, Admiral Sir Charles Madden, Bart., G.C.B., and their executive officers and port missioners for making this possible.

Alan Villiers

Chapter I

THE PRICE OF FISH

UNTIL seconds before she lurched below the sea, it had occurred to no one aboard that she would roll over. That she could do such a thing was not in their minds. They were Arctic fishermen. It was winter, blowing a screaming gale with the vicious wind picking up the surface of the sea in sheets inside the Isafjord, and flinging the freezing water over the ship. Wherever it touched the steel it turned to ice. The steel decks, the whole upper works, the forepart of the bridge were coated with a thin layer of ice, the steel railing of the upper deck bloated with the frigid stuff. It sat upon the foc's'l head in ever-increasing, silent bulk, and crept slowly up the stays. The north-east gale was blowing right into the fjord, straight from Greenland.

The *Ross Cleveland* could not anchor. She *had* to 'dodge', which means to keep going slowly, heading stubbornly into the more stubborn gale. To dodge meant to risk gathering ice. To stop meant driving ashore, to be flung upon some dreadful rocky ledge and ground to pieces. What else to do? To go outside—if they could get there—into the open sea meant dodging again out there where the sea ran in a savage, snarling froth, breaking 30 feet high. Inside the fjord, by comparison, it was almost quiet. The wind must change some time. Then she would be sheltered there.

Well, they had all 'dodged' through many such a night before and would do so again. It was part of the Arctic fisherman's life. The ice could be a menace, but it was not bad. The modern trawlers were good ships. They could take the sea's punishment. They *had* to take it. They were dry and warm inside, with the men no longer living in the plunging steel box of a forecastle-head in the bows. They lived midships now, in comfortable berths. The galley, too, could function through anything. The radio-man was secure and warm in his shack, seated among more apparatus than colleagues in many an ocean liner. The control-centre of the little bridge was warm too, with a seat for the skipper and everything to hand, including his own microphone for the voice-radio, and a clear-view glass whirling round throwing off sea and frozen spray, so that whatever was to be seen was in his view. He saw there only the storm-lashed sea and the plunging, leaping bow.

As for the men, if they could not fish they could sleep, and their capacity for this was great. They all had years of sleepless leeway to recover from their hard life, and in the snug quarters below they slept the sleep of the dead. When weather permitted and the fish were there, and they could catch them, an 18-hour watch on deck was the way of things, followed by six hours off, and this might go on until the work was done. It was not the normal watch-keeping method, of course: no one was asked to work such hours unless strictly necessary for the sake of catching the fish and getting them below. The better the fishing the harder the work. Well, there is no way round that. The men are on basic pay of some £20 a week, and the extra they earn is based on a 'poundage' rate on the fish caught. So all have incentive to get on with the fishing when they can. A trip in an Arctic trawler is not for those seeking a rest-cure: the ship must get her own cargo out of the sea with her own crew; the bigger the cargo the more work, and the more money. When fishing is poor it takes a few hours or less to clear the pounds: the watch is then free until the next haul, though the 18-hour permitted spread of their watch still stands, if wanted.

The trawlerman's days are hard enough and his life difficult, without need to exaggerate. He gets long watches enough. But a careful examination of at least the Boyd Line records shows the *average* working day per man per voyage to be about 10½ hours, and not 18. It is obvious, too, that the larger the crew the less the work for each, and the less the pay too, for the poundage from the same catch would still be divided among them all.

So nobody minded an 18-hour watch when the chance was there to profit by it. The work was skilful and exacting, of course. They knew that. To work the heavy wires that drag the trawl, drive the giant winch, see the otter-boards and all else go down clear—'shoot' the gear, drag the trawl, scoop up the fish if there, recover the trawl, empty the cod-end, gut the fish, stow them in the ice-room, shoot the gear again as quickly as possible, with often a long or short session of net repairing while it was on deck, all hands tearing at it with the wooden 'darning needles' they handled so skilfully—in, out, twist, hitch, tighten with rhythmic swing and endless motion, and the savage wind so cold it froze the moisture in their eyes.

Now they could rest, unmindful of the screaming wind still rising, the tumult of their little ship in the sea. The wind would drop again, the sea go down. They always had. And the men would come on deck again, and shoot the gear to scoop up the large fat cod, the tasty haddock and the succulent halibut, for the cargo to be hurried when the ice-room was full or fishing time was up to the swift auctions at the fish-dock in Hull. Then they would rush ashore to their little homes, to reach—so briefly—the bosoms of their families again and live not the lives of shore-side workers but to revel in the moments of home warmth and human love—so hard-won and briefly granted—with so great a joy they hated to sleep through one of them.

So it went, voyage after voyage, year after year. So it always had gone, for these Arctic trawlermen were deep sea fishermen born—off to sea at 15 in trawlers out of Hull, or Grimsby, or wherever it was their fathers and their uncles had gone before them, under far worse conditions. (Except for the weather on the Arctic banks. *That* showed no change.) Many aboard had some relative beneath the sea—not that they allowed their minds to dwell on this sort of thing. They accepted it as they and their forebears always had: accepted it and forgot it with that fatalistic optimism of the seaman and the fighting man. Death is for others, not him! So, accepting the absence of thought on such a grim subject, he can go cheerfully on.

Each was well aware that it *could* be him, at any time, for he knows that the sea grants no immunity, and modifies its malevolence for no man.

Now, for the moment, he could rest, unworried. No scream of the wild gale nor vicious rush of the wilder sea burst into his quarters. Below decks she was as tight and snug almost as a submarine. The little *Ross Cleveland*, after all, was not facing the open sea but inside Isafjord, on the north-east coast of Iceland. There she should be able to 'dodge' or lie through anything—go slowly into the gale with headway enough to steer, or lie-to with engines stopped a while in order not to go too far nor use too much fuel, remaining more or less in the same place. Her Arctic trawler hull they all regarded as the most seaworthy ship-form there could be—deep, low, magnificent in the sea, with high flared bows to thrust breaking water aside, throwing it away from the ship instead of back aboard or digging it up by the hundred-ton to fill her working decks. Minimum rigging reached aloft to hold high ice up there—the radio aerials and their spreaders, one stumpy, sturdy mast: the small stack stood abaft the compact, instrument-filled bridge. Could she ice up and fall over? Dodge to her death, overwhelmed at last in the murdering sea? Or capsize in shrieking squall of hurricane force under assault by some swift-moving mountain of wind-maddened water coming in at change of tide or over foul ground, sweeping over the bulwarks and remaining there, pinning the ship down? For high-flared bows can blow off, too, when the ship lies-to and, having driven her across the sea's trough, hold her there. And that could also be fatal.

They had dealt with the ice as well as they could, coming out on deck and smashing it with axes, crowbars, anything at hand. It is the *sudden* accumulation of heavy icing that destroys stability and overwhelms ships as it had the Hull trawlers *Lorella* and *Roderigo* north of Iceland some years before. Ice growing normally from the sea's rush may be relied upon to increase with reasonable slowness, allowing the deck-crew time to cope.

So it was this night: so it had been through four previous nights. Wind and sea were sufficient enemies.

The *Ross Cleveland* dodged and lay, survived and hoped. Inside the warmed bridge, full of fish-loops, self-steering, electronic navigational devices plotting her stumbling track with steady precision, Skipper Philip Gay kept in touch through his powerful voice radio with other Hull trawlers, especially his friend Len Whurr, also dodging in the Isafjord, quite close.

He hoped the wind would go down, of course, so that he could fish again. His living, and all the crew's, was in his catch of fish. While he was forced to dodge the ship consumed her fuel, the ice in the ice-room slowly consumed itself—it was makeshift preservation at best—and the fish, in time, could begin to deteriorate. There was no great number in the fish room then anyway: its storage was good for two or three weeks at best, and the gale could blow for half that—or more. Meanwhile the steady consumption of her fuel and the failure to add to her cargo must, in time, affect the stability of the sturdy trawler: alone of all ships, a trawler must live in the sea and take her cargo from it, too. Now the sea's build-up was flinging the ship about and the hurricane gusts trying to blow her over as she lay-to. If this was too long-continued, just how much righting-moment—that vital force dependent on assured stability which kept her upright and brought her straight again when she rolled—might be left?

The Skipper had no instruments to tell him that. Nobody had. At that stage, it would have made no difference anyway. He could have done nothing about it.

If only the infernal wind would go round, back or veer anywhere, and no longer scream directly into Isafjord! Then one side or other of the long north-east south-west funnel that was the fjord would give a lee. He could get an anchor down until the weather eased enough to carry on fishing.

But the wind got up to an official Force 11—near hurricane strength—with no sign whatever of easing or of changing direction.

Lying-to now, she was sluggish to come up again into the wind. At times, she 'hung' on her heavier rolls, sluggish to lift again, to regain and keep some semblance of the upright. Heading into the wind, she did better: but now she was almost in the trough and showed reluctance to push her bow back into the eye of the gale. Deep in the water aft, the big propeller thrashed. The rudder was hard over to force her hull round, engines half-ahead. Still she hung.

Full ahead! The telegraph changed.

She began to swing. At that moment, a tremendous sea struck her on the weather bow. She lurched, staggered, fell from the blow, filled her decks with the sea, and the sea pinned her down though it gushed from all the washports. Like a wounded thing she fell off again, and more sea came, and she would not face the force of the wind. Now she was fighting to recover balance, to stand up. She could not fight the weight of sea and the wind's power arrayed in such terrific strength together while too much shipped water held one side down.

Tense now, Skipper Gay clung to the wheel-base, the bridge and its instruments at a crazy angle. He still had his microphone in one hand.

"Help us, Len," he spoke into it now, in a quiet, clear voice. "Help us: she's going over . . ."

But Skipper Len Whurr could not help. He was fighting for his own ship's life. Neither with his ship nor her one boat could he come to his friend's aid in those terrible conditions. He could not launch the boat. Even if he could, no one could handle it in such weather. He had life-rafts: they were for abandoning ship only. You took to them when the ship left you, and prayed, for you had no control over them in such conditions unless you were close-by to windward, and by God's grace they drifted instantly in the one right direction.

"We saw the lights inside the bridge stand vertical on the sea," he recalled later. "Then suddenly they went out."

Skipper Whurr hoped that Skipper Gay, or his men, had managed to get her life-rafts away. He kept vigilant lookout for them, his searchlights stabbing into the black and screaming storm. But the freezing spindrift howled forever in his face. He saw nothing at all, heard only the wind.

The *Ross Cleveland* did in fact get one life-raft away. In it was the Mate, Mr. Harry Eddom, of Hull, aged 26.

Mate Eddom was on deck when she went. He had been chipping ice from the radar apparatus on the bridge. That was why he was in the full fisherman's oilskins which became his survival kit.

"When she began to go she went in six seconds," he said. "It wasn't the ice. It was the wind and the sea."

He was pitched off by the capsizing roll. By great good fortune two of his shipmates, Bos'n Hewitt and spare-hand Barry Rogers, who had rushed up from below in their underclothes, had at least got one of the rafts to the state where it slipped into the sea. They saw their Mate, and hauled him aboard. These trawler life-rafts are the same pattern as air-rescue rafts and are a splendidly designed, quick acting device to save men's lives. They have canopies which inflate quickly and act like windproof and waterproof floating tents, inside which survivors may raise the temperature by the heat of their own bodies.

But the bos'n and the spare hand, having dashed up at a second's notice from below, had little body heat, no survival suits on, or much else. Their bodies were cold and had no chance to warm up.

When the raft miraculously drifted ashore in Isafjordur, only the Mate was alive.

(*Note:* That storm really was a terror. According to the Chief of the Icelandic Gunboats Service, if it had gone on for three-quarters of an hour longer than it did, at least another forty vessels would have been lost including their own patrol gunboat *Odin.*)

Chapter 2

HAZARDOUS CALLING

MATE Eddom was not just the only survivor from the *Ross Cleveland* that terrible night. Around that same time, two others of Hull's Arctic trawler fleet—the distant waters ships *Kingston Peridot* (658 tons) and *St. Romanus* (600 tons) were lost with all hands. They disappeared, and no man knows what happened to them. An empty life-raft was found from the *St. Romanus*, an oil-slick that may have come from the *Peridot*, but nothing ever of their forty men and boys. They sailed, and they did not return. Mate Eddom was the sole survivor from three crews.

Some years earlier, two other fine Hull distant-water trawlers, the *Roderigo* (800 tons, 21 men) and the *Lorella* (600 tons, 20 men) were lost when trying to fish by winter on the north-of-Iceland grounds: but all fishermen knew where they were and what happened to them. They 'dodged' to their deaths in an appalling Greenland winter gale, too: but the ice got them before the sea did. They iced up and they rolled over: nobody got away. In both cases, their courageous Skippers—George Coverdale of the *Roderigo*, Steve Blackshaw of the *Lorella*, both Hull men—were heard to go, announcing their fate as their iced-up trawlers lost all righting moment and rolled into the storm-locked sea. For the two skippers were friends, and they were speaking to each other by R/T on the trawlerman's waveband through their merciless trial.

"Going over to starboard," brother skippers in the trawlers *Lancella* and *Imperialist* heard the quiet voice of Steve Blackshaw say: and an hour or two later "I am going over now. I cannot abandon ship. Mayday!"

Then it was George Coverdale's turn—almost the same words, spoken with no trace of panic, the same call of Mayday!* Mayday! as his ship went down.

One difficulty about deep sea trawling is that the longer a vessel is forced to keep the sea dodging in a severe gale and unable to fish, the worse her stability problem can become. She consumes fuel and replaces it with nothing. Alone among ships, she must take her cargo from the sea, in the sea, and her people must work her trawling gear and gut the fish on her open decks (though this is no longer so in the stern trawlers). They must open hatches to stow the cleaned fish in the ice-room down below. They are small hatches, very strong, designed by men well aware of the need to make them swiftly water-tight. But they are openings: and the crew are there to use them.

All sea fishing can be dangerous—seamen know that and fishermen live with it. But the distant-water Arctic side trawler—the now old-fashioned type which tows her

---

*Mayday, chosen because of the difficulty of misunderstanding the short, clear word, is the spoken SOS of international Radio telephony, for ships and aircraft.

huge trawl by means of wire warps led to the sea bed from her starboard side—leads one of the most hazardous lives in the world. Of all men going to sea, cold statistics show her people to be at greatest risk. It has been computed statistically* that the trawler deckhand works with a three times greater risk of death than the coal-face miner, who is on risk for only a 7½-hour day anyway. Skippers and mates, more exposed and working longer hours at greater stress, are worse off, for the odds against them are *twenty times greater than those of any other industrial risk* in the world. Above all, the distant-waters trawlerman, habitually working for his living by hazardously harvesting the grim waters beyond the sea lanes, summer and winter, faces heavy risk not only from the possibility of total loss of his ship but from personal accidents while going about his work on board. For trawlers have low freeboard the better to scoop up their nets from the sea: it is easy to go over the side. They have limited endurance, particularly with the old ice-box method of keeping their fish fresh for market. Therefore the stress is on their men: the tendency is to work while there are fish, for time presses.

The men work with sharp knives. They stand in slippery fish-pounds, hour after hour. The foredeck is a fresh-fish catching-and-cleaning plant: heavy wires writhe from the great trawl-winch from which the trawl is operated. Giant, steel-shod vanes called otter-boards hold the net open on the sea bed: these come aboard with each haul. The trawlerman runs a risk of being minced in the

---

*See TRAWLER SAFETY, Report of the Committee of Inquiry: Chairman Admiral Sir Deric Holland-Martin. H.M.S.O. Cd. 4114, 1969.

trawl warps (now much better protected than even a few years ago but still dangerous), mashed by the otter boards or pitched overboard while tending them, flung into the sea, pitched on his head into a steel-cornered hatch on the slippery, gut-strewn working deck where the sea sweeps always but never enough to wash away all slime. If he gets in the sea he has little chance of being rescued unless the weather is good, or another trawler in company can get to him: for trawlers have one ship's boat often little used, and life-rafts for life saving when the ship has sunk. Arctic fishermen are heavily clad when they work on deck.

Of Britain's 7,500 trawlermen fishing the more distant waters, out of the ports of Hull, Grimsby, Lowestoft and Aberdeen—each with a hundred or so such vessels—an average of well over a thousand men a year suffer serious accidents. Between the years 1957 and 1966, nine such trawlers were lost by foundering or capsize. Three more went with no trace at all. Another forty-four were total losses through stranding—sometimes, perhaps, the result of fatigue upon the bridge—and twelve by collisions.

So it goes. So it always has gone, in this hazardous business of professional sea fishing. As the ships become larger and of greater endurance, they have gone farther afield—to steam 2,000 miles and more before shooting a trawl is commonplace for the distant water men, which now habitually fish the Davis Straits off the west of Greenland, off Labrador, on the Newfoundland Banks, north of Iceland, Norway, and Siberia, round Bear Island. Well-known former banks—some of them developed by the British, pioneers in the field of distant water trawling—have been withdrawn from the international field by ever-extending national sea-limits, electronically patrolled and rigidly enforced. In areas of the sea once thought severely hazardous where (because the sea lanes do not pass that way) no merchantman is seen, the little trawler is at work, ice permitting, all the year round. In seas the old-time Polar explorer knew as grim and desperate and was liberally bemedalled for once having braved them, the trawlerman accepts the necessity for his 18-hour watch when he is on fish, come what may short of hurricane.

Of course, from the trawlerman's point of view, it isn't all that bad. After all, the vast majority of voyages are successful. Most trawlermen die in bed, though on the average perhaps at an earlier age than followers of other callings. The rewards can be good, often very good. At sea there is a wonderful camaraderie. A man may savour brief tastes of the shore life with a greater joy than the land-dweller who is always there. His is an adventurous life, with its own values. Women appreciate him, though loath to be so often left.

The East Coast of England and particularly the Humber, with its ports of Hull and Grimsby, has a long tradition of off-shore fishing. Toughness of these fishermen goes back at least a hundred years. Ever since those splendid seamen the Brixham fishermen, probing far afield as usual, discovered the 'mine' of fat flat fish on the Dogger in 1850 (the 'Silver Pits') and the fleets of smacks based themselves on the Humber the better to use the find, the fisheries have taken their toll. The smack-fleets had to ride out the seas no matter what came for months on end, but they could use their unwieldy 'beam' trawls only in the right conditions. These were nets with cod-ends drawn along on a large wooden spar slung abeam, dependent for its motion along the sea-bed on a sufficient breeze in the smack's sails. No wind, no fish (except with lines): too much wind, no fishing either. So, as long as his smack—a beautiful seaworthy little vessel, ketch-rigged, easy to snug down, deep-keeled to live in the nasty shallow seas, buoyant and graceful to defy them—could remain afloat, the smacksman was all right.

Hull had been an Arctic whaling port since the 16th century and sent its first auxiliary steam whaler to the Greenland grounds in 1857.

Within a year, in 1858, the Lincolnshire port of Grimsby near the Humber's mouth had built the little steam trawler *Corkscrew*, well-named for her motion on the sea. Ten years later, Hull's last Greenland whaler, the *Diana*, was wrecked at the mouth of the Humber. The Greenland whale fisheries were a suitable background for seamen of the hardihood and competence to handle the port's hundreds of smacks. From these tough beginnings the traditions and the cheerful acceptance of his way of life (no matter the cost) have sprung. Men whose grandfathers were Arctic harpooneers, chasing giant whales from small boats, could take whatever hardships and dangers the sea-fisheries might hold in their sea-booted stride. So they did, and they still do. Hull's first steam trawler, a small vessel named the *Zodiac*, was built in 1883 —almost 80 years after a primitive steamship had first operated in Britain. These were conservative men, not given to the over-quick fostering of new-fangled ideas. But quickly they saw that steam-trawling made sense and took it up in a big way. Hull's smack-fleet declined while the powered trawlers flourished: but the traditions remained the same.

Fishing banks were extended, the otter trawl introduced (1895)—and the Hull Fishermen's Widows and Orphans' Relief Fund along with it—and, as far back as 1905, the White Sea grounds were discovered. The White Sea, to British trawlermen, is a general term for the waters beyond the north of Norway, Finland, and towards Novaya Zemlya. Many prolific fishing grounds extend past Norway's North Cape, off the Murmansk coast eastwards and northwards to within sight (if visible) of Bear Island.

But the smacksmen long retained their conservatism. Their running costs were low: many smacks were individually owned and family-sailed: they offered a meagre living at best, but they were a way of life as well as a means of supporting a little home. They cost little, required simple, inexpensive maintenance, lasted for years.

Distant waters meant large ships with large power, requiring large investments—first thousands, soon scores and then hundreds of thousands of pounds. Large investment demands large earnings. Fishermen have traditionally been paid by some form of shares. Unlike merchant seamen working for a wage in a ship paid for carrying goods on a defined route, the fisherman may have no return—nor the owner—until he has won a saleable cargo from the sea *and* landed it while still in saleable condition. In a distant water trawler, he has little more than three weeks port-to-port to do it in. The larger the crew the less the share for each man, obviously: the crew for so hard-worked a ship has long been accepted as minimal. Hence acceptance also of the necessity for an occasional 18-hour watch for the deckhand, the long hours for Skippers and mates, and so much else in this courageous industry.

So the stress is inevitably on all of them, as it never was in the comparatively small smacks: and stress is an appalling ship-mate, the nastiest and most active accident-breeder known to seafaring man.

If distant-waters trawling in the now old-fashioned side trawler, instead of being long established and accepted, were a new industry, it is difficult to imagine how it would find recruits: for it is manned by brave men born and raised in it, accepting it from early childhood as a matter of course. Fishermen are not drawn from the usual run of seamen. To them their ship is a means of catching marketable fish where they may be found. To gain a living in this way they must accept the risks of their calling. Nor does it seem to them that these might be regarded as excessive, for these are in the main committed men, from experienced families, leading the way of life they thoroughly well know and with which they are content.

Rehousing the fishing community which used to live round the Hessle Road, which has followed some sensible replanning in Hull, is changing the old outlook of these formerly well-knit folk: the old committal to fishing is no longer so strong, old-timers tell me. But it is still there.

Chapter 3

TO DISTANT WATERS

They looked, at a casual glance, anything but seamen, those pale, slight young men coming slowly through the dockside litter and climbing wearily aboard the distant-waters trawler *Arctic Vandal*, bound for the White Sea. Cloth caps on fair heads, shore-going suits, shoes, ties and all the rest, they might have been clerks or factory-hands going somewhere to punch the clock. But there was a lean litheness about them, too, a spring in their step (though they looked tired), a sure skill in the way they picked the shortest path through the litter of broken otter boards, elderly trawl warps, ancient gangways, empty metal fish-kits fallen from careless motor-lorries and the mess of mooring lines from a huddle of adjacent trawlers, and vaulted over the low bulwarks in the waist of their own, dragging their sea kit with them. The tide was early that morning—it always is, or serves in the middle of the night when the day's Hull distant-water trawlers put to sea.

The fish dock is locked from the river, very naturally: if it weren't it wouldn't be a dock, for the water would spill out of it with every ebbing tide. So the trawlers may go to sea only at or near high tide when the water level inside and outside is about the same and the heavy lock gates can be safely open. In the old days, a group of the little steam trawlers, noses to the quay, gave an air of impatience with smoke belching from high funnels and plumes of steam at the high valves. Now with diesel motors or oil-fired boilers there is no smoke and little steam: fish-hatches are battened down with their load of ice other men have put aboard: the big trawl and its gear lie below ready for their work where the trawlermen repaired and stowed them, homeward-bound from the grounds last voyage. Now the Chief starts his engines, lines are taken aboard, the Skipper gives his usual exhibition of that marvellous shiphandling skill all Skippers must be born with—a flurry of frothy brown water from brief prop-thrash below the counter and she is away, gliding down the dark dock as if glad to leave its litter behind. No fuss, no shouting, no whistles blowing, for these are ship-handlers of such perfection there just seems nothing to it.

The little *Arctic Vandal* of Tom Boyd's *Arctic* fleet out of Hull is bound for yet another hard voyage: the tide serves, and she goes. Nobody looks at so ordinary a scene. Her people are aboard (one or two joining at the lock gates, where they had been replenishing sea stocks of their Arctic gear, at the special owners' co-op store placed conveniently there, offering the best of gear at the lowest prices: gloves, sea-boots, expensive oiled-silk outer suits are costly and must stand heavy, demanding wear).

Everything she needs is aboard, though the crew have not been near her for the couple of days she has spent in dock. They rush ashore as soon as she is at her landing berth—others land the fish, move her about the dock, store her, ice her, fuel her, maintain engines, bridge and radio equipment and gear as necessary. The time in port is precious to the trawlermen and they are not required aboard again until their ship is about to sail.

The familiar scene of the big St. Andrew's fish dock does not interest them this morning, with its fleet of trawlers of the old side-dragging type and its big new stern freezers. To me, these lack the sea-panther grace of the good-looking old-timers, long and low, but I know they have great advantages of their own. I looked at them with interest as we slipped past, third in the line of distant-waters men going that morning. I noted their useful high freeboard, and massive hulls. I knew of their immense endurance, their ability to keep the sea for months, to fish in any weather and clean the catch in the shelter of a factory 'tween-deck, freezing the cleaned fish instantly at prime quality and holding them so indefinitely—all this, and a simpler handling of the cumbersome trawl, to say nothing of roomier and more comfortable quarters for the crews, seemed to me to add up to considerable advantages for the modern big fellows. But I note that I am the only one aboard to glance twice in their direction, to note the points of the *Cassio*, *Lord Nelson*, *Ross Implacable*, *Prunella* as we slip by.

"They stay at sea too long," said Skipper Bernard Spicer when we'd reached the river where a big one was swinging compasses. She was a utilitarian job. She looked the factory ship she was—square-cut and clumsy-masted with two sets of heavy goal posts bolt upright to work the stern trawl, and a light mast farther for'ard with a pronounced rake. Twin stacks were stumpy and upright, too.

"They keep the sea for seven or eight weeks. My wife doesn't care about that. She likes me home more often. A couple of days every three weeks is better than one week in seven or eight.

"But she might have to put up with her husband in the big ships yet. There hasn't been a side-winder built for Hull since 1962. Maybe even the wet fish trade is on the way out."

I'd learn more later, for I was to be aboard at least for three weeks and switch to one of Tom Boyd's stern-freezers in Arctic waters, if the chance came. He had three of them as well as ten like the *Vandal*. All the hundred or so trawlers out of Hull are distant waters ships, and in recent years there has been a marked swing to the more modern type because of their obvious advantages. But the smaller of them costs three times as much as a side-winder like the *Vandal*.

The side-winders' people obviously preferred the way of life they knew: so, apparently, did their families.

I would have a chance to watch these new-type ships at work during the coming voyage, for they fished the White Sea grounds and Bear Island too. In the meantime, the lads who had come aboard glumly enough soon picked up their spirits. They were a cheerful lot, even going to sea that grey, early morning. They were obviously all friends and acquaintances of longstanding from 62-year-old Brian Taylor to Deckhand learner Steve Collier and Pete Somerville the juvenile galley-boy, aged barely 16.

'Deckie-Learners' nowadays must serve in the galley for their first few voyages and must not begin at sea in the tough winter months. Young Pete—born to trawling, like all the others—could not wait to get on deck. He was cabin-boy, crew's messman, and sort of second cook. As for Steve, he was practically a spare hand already. (Aboard a trawler a spare hand is by no means 'spare'—not in any sense. In fact he is a fully qualified fisherman though not rated as third hand, for these are under-officers and require further qualifications.) Both Steve and Pete had already done courses at Hull's Nautical College, and Steve had also served in the school-ship.

"Nowadays lads don't go on deck until they're a reasonable size, have made at least three trips in the galley, and have a D/L's ticket from the school," Tom Boyd told me. "We have 20 deckie-learners, and 20 galley-boys. Thirteen of each are at sea, the others at school."

There is plenty to learn. The previous voyage, the deckie-learner had gone overboard—just how nobody knew.

"I've four uncles in the sea," said Steve quietly. "All from trawlers."

I thought he'd said *at* sea, at first.

"No, *in*. For ever." He made it clear. This seemed a lot of relatives to lose, even in that hard life, but I gathered that he had many other relatives who still flourished in it and a grandfather, dead recently at an advanced age, who'd sailed in the engineless smacks. It was his yarns about smacks which had first sent young Steve to sea, though the *Arctic Vandal* was as like an old North Sea sailing smack as a modern automobile resembles a farm dray.

Beyond Spurn Head the ship hurried on her headstrong way, bucking, leaping, flinging the heavy sprays over herself as she hurried to the north. The nearest grounds where she could usefully wet her trawl were beyond the North Cape of Norway, over 1500 miles away, and the wind and sea were right on the bow. This was the usual thing. She was well battened down, and her gear—as is the usual custom—had been made ready on the last homewards run. For a day or two, there was no need for anyone to go on deck. The little bridge was water-tight and self-contained with all her navigational and handling needs up there and, abaft that, the snug, secure radio-room, packed with more equipment than many an ocean liner ten times her size. The crew's quarters though of course not battened down were compact and weather-proof too, a self-contained and self-sufficient steel house, reached at sea by sheltered steel doors opening aft (and so out of the wind) and a circular escape hatch, to the upper deck. Here were the engine-room, gyro-room, galley, store rooms, mess rooms, and quarters, heated and ventilated, with the crew berthed in single or two-berth cabins, snug and secure from the sea. From this built-in 'house' there was ready access to bridge (from there to the upper deck), radio-shack, engine-room, everything. The quarters were totally enclosed, without ports on the sleeping deck below. You could not see the sea from down there. I gathered the trawlermen liked it like that. They saw enough of the sea.

Bos'n Ted Newman gave me his cabin, a warm, roomy place with good head-room and two bunks, good floor space for a man to wrestle himself into and out of his fishing-gear, a couple of large orange-coloured life-jackets, and a couple of wardrobes full of fishing gear—cordage, synthetic stuffs for net repairing and wooden bobbins to use on that job, shackles of assorted sizes, and so on. As a good bos'n, Ted had a large hoard of such gear in the instant-use locker he'd converted his clothes-hanging space into: and, like a good bos'n too, he left it all there. He knew where what he wanted was, and this is a large part of the secret of good bos'ns—seamen worth their weight in gold anywhere, but especially in that impatient, sea-fighting game-cock that is a deep sea trawler.

I settled in here with no difficulty, while the little ship hurled herself along towards the fishing grounds. Around me the mariners slept the sleep of the just and the secure in spirit, resting while they could. But only for the first day. Soon the mess-room rang with their cheerful laughter, for they were happy men with their feet on the earth though a trawler was their normal home and a fishing voyage their normal habitat. They spoke of a 'trip' to port—not to sea. Sea was where their work and lives were.

Up on the compact bridge, Skipper Spicer had charge. He was Mate, usually, under Skipper George Kent: George had been fishing very well over many voyages and now was taking a holiday in Spain with his family. Bernie looked older than his 32 years (for skippers and mates can age fast in Arctic fishing). He too, came from a fishing family (there was a Spicer lost in the *Lorella*) and had been in trawlers since his early 'teens. He'd been skipper with another company, but amalgamations and the trend to larger groups with fewer ships sent him back to Mate again. This was his first chance to go Skipper with Boyd's. All skippers, well aware that, to some extent anyway, the taking of good cargoes is considered as up to their flair and skill, are under stress, and a first chance is important. In a trawler the skipper is all-important: indeed, the voyage is in his hands and that indefinable quality called luck or flair inevitably comes into it.

But 'flair' is acquired by infinite ability to comprehend the data supplied by instruments, based on years of practical skill and long knowledge of the wide scattered grounds and the habits of schools of marketable fish: nobody can begin accumulating much of this until he *is* a skipper. It is not taught. How the sets run over a hundred banks below the Arctic Ocean, the vagaries of the codfish on the Grand Banks of Newfoundland, on Storre Hellefisk off Greenland or Nortons Ground off the Murmansk coast and beyond the 12-mile limit off Skomvaer—above all, where to find a pay-load of fish *every* voyage . . . this is part only of the priceless store of the accumulated know-how the distant-waters skipper needs. So he sails for hard years as deckie-learner, spare hand, bos'n, mate—down on deck doing the donkey work, in the pounds cleaning fish, in the fish-room carefully stowing them, while the Skipper, alone, fishes from the bridge. So how does one go about the accumulation of the essential knowledge? It isn't born in anyone: but common sense, *sea* sense, inexhaustible energy and unquenchable courage—these qualities are, and they and a quiet, steady, mind are a firm foundation.

These things Bernie had, in good measure. He was a cheerful young man, slim and pale like the others, come aboard in a lounge suit, with well-trimmed hair. There was little or nothing about him ashore by which the landsman might judge the arduous nature of his calling. Most landsmen know nothing about it anyway.

Now he spent all day and much of the night up there on the bridge, with a watch-keeping hand as stand-by helmsman and lookout, while the automatic pilot kept the ship plunging into the head sea right on course regardless of the heavy seas which clumped and broke over the fore deck and sent sprays continually smashing at the bridge windows. She shuddered, she trembled, she bucked, she shook herself like a wet old sea-dog. But northwards she ran with her lights burning brightly, her hands sleeping the sleep of the exhausted for they soon would be exhausted again, her diesel thumping away full-out, her young skipper watching over her while he hurried her towards the White Sea grounds, and kept his hopes and his thoughts to himself.

Chapter 4

IN THE *ARCTIC VANDAL*

THE *Arctic Vandal* sailed that voyage on August 28, 1969. It was the first of September before she began to fish, 1400 miles away, on a bank off the North Cape of Norway. There was news of Russian naval manoeuvres somewhere in the Barendz Sea area, and the U.S.S.R. authorities had indicated that it was better for the moment not to fish there. This had been done in no peremptory way, I gathered. A trawler or two had been boarded, and areas free from war-ships indicated to them by officers speaking excellent English.

"We get on all right with the Russians. We often fish among them," said our Skipper. "I wish we always got along as well with the Icelanders."

He was speaking only of factual fishing operations, of course. Like all seamen, the Icelanders share that fine spirit of international co-operation for which working seamen are noted. It isn't fishermen who fix limits, or patrol them either. Many a British and other trawler has been grateful to the helping hand of the Icelandic protection vessel *Odin*, though a few might have found her presence awkward on other occasions.

Well, here we were, on fishable grounds at last. What was different about this particular stretch of cold grey water tumbling restlessly in all directions, I couldn't see, or why the Skipper chose to begin there rather than elsewhere I had little idea. He had instruments, restless devices which 'pinged' on fish swimming in the waters near the vessel, some keeping a graph (with blobs for fish) with scratchy pens swinging endlessly in busy circles, others showing blobs of gyrating green light. Yet another strange device told him exactly where he was, keeping a graph of that and of his course, so that if he got above a good school of fish he could mark the place, go back and have two 'goes' at it. It wasn't as easy as all that, of course—nothing in fishing ever is. You had to read the graphs and the blobs correctly. The *Vandal*'s trawl, with its great mouth rolling along the sea bed on a wide line of steel bobbins, trapped demersal—bottom-dwelling—fish only (cod, haddock, halibut, plaice and so on): the instruments 'blobbed' the lot. Fish swimming a few feet above the bottom could perhaps avoid the net's fatal embrace.

Now they were on the bottom. There were other indications. Other Hull and Grimsby trawlers were about and the bridge receiver, at full blast to miss nothing, broadcast the Yorkshire accents of a dozen loquacious skippers talking away about everything but fish. Bernie knew many of those voices all too well. He could detect the state of their fishing no matter what they said. It was how they said it that gave them away. Sometimes one did give a bit of 'news'—'50 baskets last haul', or something like that when he'd probably got 150—and sometimes a few words of truth burst through the chatter—in code. Bernie knew the code.

Anyway, part of the fleet was there—and staying there. Good enough. And the blobs and graphs were sufficiently encouraging.

Away with the trawl!

Not that anyone gave such an order. Nobody said anything. I watched, well out of the way of skull-bashing otter-boards and writhing warps. Those men knew their work so well, no orders were necessary. Despite the bad weather, they'd been ready for days. To the layman, the business of getting a big trawl away clear and ship-shape to tow from the starboard side (the only side rigged for it nowadays) seems extremely difficult and complicated. From a large rolled-up net stowed along the length of the starboard scuppers, a lot of heavy wire on an enormous and apparently complicated winch on the after end of the foredeck and leading through steel piping and over horizontal, vertical, and slanted steel guide-wheels to a couple of steel gallows leaning outboard, each with a heavy iron-shod wooden platform chained to it—the otter-boards, which are in effect vanes designed to keep the trawl mouth properly extended and fully open—from all this apparent complication somehow to get the lot clear overboard and doing its work along the bottom in the only way it can, struck me with wonder and admiration every time I saw it done, though the trawl was 'shot' six times a day for most of the following fortnight.

Obviously, the net had to stream clear so that it would set itself properly on the bottom and not roll itself in a bunch of knots behind crossed wires and fouled-up otter boards. Equally obviously, the lot had to avoid the propeller, or the ship wouldn't go at all. The otter-boards had to set themselves upright about 80 feet apart, towed abreast of each other at the end of wire warps which led from different parts of the same side of the ship. The bobbins were required to roll along evenly across the sea bed, which was not always possible. The sea bed is uneven like the surface of the land, often rock-strewn, sometimes studded with rusted steel bones of too many wrecks. Every single piece of the gear—the special Dan Leno bobbins, rubber wing bobbins, anti-chafe hides towing below the cod-end to prevent that vital fish-trap from being burst or chafed into holes too easily—*everything* had to be precisely where it belonged. Of course, all is set up correctly and kept that way whenever it is aboard, but the ham-handed or the inexpert could snarl the lot in the twinkling of an eye.

I watched from my vantage point between the ready-use life-rafts on the upper deck—one jerk of a wooden toggle and they were instantly freed when needed—while the orange-oilskinned experts, mindful of the need for those opposites speed and care, went about their work under Skipper Bernie's ever-watchful eye. Otter-boards clanged briefly, bobbins brushed the ship's rail noisily for a moment, the cadaverous countenance of the Mate looked more determined than ever down there on deck among the wires, leaping over the fences round the pounds, the guards on the warps. The alert young Bos'n—swift, competent, assured, lithe in his movements, always instantly at the spot where something threatened to go wrong before it went wrong—backed him up. So did old Brian Clark and Bert Bilby, Eric Everson and the rest of them down to Steve the Deckie-Learner, while acting galley-boy Pete looked enviously on. The trawl streamed to wind'ard well away from the ship, which was blown away from it and not on it.

Old Brian and a young trawlerman stood by the great winch manipulating controls to keep the tension on the warps just right. The winch was immediately below the bridge windows. Bernie had a perfect view of the whole fore-deck and along the side. The starboard side of the ship was open: he could see what happened through his side windows, or step out (if he wanted) on to the upper deck. Everything was very well arranged and organised, the fruits of long evolution. This trawling, I knew, went back to the sailing days of the North Sea (and Irish, Scots, Welsh, Flemish, Scandinavian, German) smacks which sailed along with their trawls extended by wooden booms, and developed quickly when power came. As for that, both Brixham in Devon and Barking in Essex claim that some sort of trawling was going on out of these ports in the 17th century and earlier.

The trawl and its gear looked very complicated to me. How did anyone know what was happening on the bottom? You couldn't see two feet through that surly sea.

"There have been experiments with T.V. cameras towed down there to watch. One trouble is light," I was told. "You'd have to use too much power to light the sea-bed for the camera to send up a useful image. Too much power for the camera too, *and* expense. No, we can feel what's happening, see how the warps trend. We get to know quite a bit about the bottom on most of the grounds we work. We just know when things are going right, or aren't."

I could appreciate that. After all, suddenly to find oneself on the poop of a full-rigged ship in a hard gale, without experience, would perhaps be somewhat alarming. But the varying sound of the wind in the sails and the rigging, the motion of the ship, the amount of difficulty in steering, the feel of the gale on one's face—these direct the sailing-ship master in the darkest night and foulest gale. He was brought up in that business. He doesn't suddenly take over.

So the trawl was down, and the *Vandal* settled to a four-knot tow while Bernie watched the fish-loops and all the instruments, listened to the chatter on the air which now was much louder for we could see the lights of other trawlers, and now and again quaffed a mug of strong, brown tea. Below, the fishermen attacked an evening meal—rich, fresh meat pie, boiled Yorkshire roly-poly, thick vegetable soup (no fish as yet, for we hadn't caught any) and the equally thick tea—not taking all their heavy outer gear off, as is the way with men alert for a call. The radio blared out world news depressing and oddly inconsequential there. Deep aft the powerful propeller throbbed to the rhythm of the dancing rods of the diesel engine. Chief Engineer Ken Blundell—like all the rest born into fishing, though he'd been a paratrooper in the British Army once—kept an ear wide open to both sounds.

"I don't know," he said, "why anyone should ever come to sea in a trawler to see what goes on. All that goes on is hard work."

He grinned.

It was strange how the noisy sea birds gathered around when they saw the fishing lights go on, as if they knew. Perhaps the fish knew too, but if they did their reaction was different. When after three hours on the bottom the trawl was hauled, the cod-end held a mixed bag of a few baskets of fish hardly worth cleaning.

The net was damaged, too, despite its protective anti-chafe covering of expensive hides. The fish would have to wait. First the cod-end must be repaired. It was taken aboard, the damaged section stretched along the deck, and the eight men on night-long watch began at once to repair it as if their lives depended on getting it back into the water with the utmost possible speed. The ship was well prepared for such accidents. There were ready-use patches of the tough artificial fibre (much stronger than cordage) ready to fit most holes. A selection of these the lads thrust into place and 'stitched' away like mad with the big wooden bobbins. In, out, hitch, haul–in, out, hitch, haul–and so on *ad infinitum* with astonishing speed and sure skill, though some stood in fish and all in water on the open fore-deck, and it was mighty cold up there a hundred miles north of the North Cape of Norway.

So sure was their touch and certain their skill, I wondered whether the trawlermen had been taking net-mending lessons from Dora Swift, champion hand net-maker at the Cosalt plant back in Hull. Dora is one of the very few women experts still in that business (at which all fishermen's womenfolk and most of their children once excelled, to help their menfolk make a living): she works like a precision machine. But she stands in the shelter of a warmed room where the Arctic wind and the storm-lashed whip of the spray cannot get at her, and she works regular hours making new nets, not hurriedly patching up old.

Up on the bridge Bernie, philosophical as ever, watched the work, waiting for the trawl to be ready to shoot again. Behind the bridge, in his instrument-packed radio shack, Operator Norman Willis listened for all and any scrap of information relevant in any way to fishing. A clear, firm voice in Norwegian was intoning a weather forecast for fishermen from off Lofoten northwards—all bad. Norman got that down, passed it to Bernie. It was quiet inside the shack, with the local chart open on the table for it was chart-house and radio-room combined. Here Norman put in prodigious hours, like everyone else, though in fact he was not bound to.

Bernie's own R/T set on the bridge chattered away, with the same Yorkie accents and a touch of Lincolnshire now and again. The skipper of the *Kingston Amber* was on again. There was talk of a haul bringing up 200 baskets—a very satisfactory catch to be going on with.

"That'll be Neville," said Bernie. "He just has to shoot his trawl anywhere and the fish fill it."

A very desirable quality—and, of course, quite unknown—for any trawler skipper to have, or even to be believed to have. We were by this time not far from the

Kingston Amber. I noted from the records that she had landed over £13,000 worth of fish from her last voyage, a 19-day run to the White Sea grounds: the *Vandal*, a smaller ship, had done very well too, with over £12,000 worth after 15 days fishing. I wished the same good fishing both for Neville and ourselves this voyage, as I looked out over the increasing fleet of trawlers around us there, fishing lights dancing from their rolling village in the sea.

This is where they live. No wonder young Pete Somerville the galley boy went off to join them as soon as he could, from the little home in Hull full from childhood of his kinsfolk—the Arctic fishermen from the North Cape and far beyond, the White Sea and Greenland, Bear Island and the north of Iceland, to whom the home was a temporary dormitory where they rested a day or two before going back to sea.

Ah, there was Neville on the air again—another 200 baskets. And our trawl ready once more to shoot.

Chapter 5

TOM BOYD OF BOYD LINE

THE second haul got some prime fish—30 baskets or so, enough to show that good fish were there but not enough to make much impression in the fish room. Bernie shot again: the other trawlers were still about. There had been plenty of good fish on this bank the previous voyage. George Kent had scooped them up until the whole foredeck ran full, and the chaps worked and worked and worked. That whole voyage had taken only 15 days, port to port: but now it was already 13 days since the *Vandal* had filled and left—five days homewards, three nights in port, five days back again. There were many other trawlers on the bank. They came and went. They couldn't all keep dragging over the same ground and catch fish.

It is a strain on a man, to give him a chance as skipper only when the regular skipper takes leave, but no opportunity to take charge at any other time. How is he going to acquire the infinite know-how of so difficult and expensive a business? He cannot learn it all from understudying his skipper. He isn't there as an understudy.

He's Mate, a full-time job plus: just one of his responsibilities is the grading, stowage and state of the catch in the fish-room, a rather vital matter that keeps him largely off the deck while gutting. He can *see* what his skipper does—and there were few if any better than George Kent—but he couldn't have much clue *why* he did what he did, on what particular line of reasoning or just plain 'hunch' plus long-accumulated, slowly-absorbed-from-early-youth know-how George acted. George shot the gear and the fish were there—well, more or less, as they were said to be for Old Neville. They couldn't smell fish. They hadn't any special 'instinct'. For what has man born of woman in common with fish?

They'd both been mates for years, too. They'd all come up the hard way. But the older men had got their chance when the stress was not quite so great. Ships hadn't cost so much to build or to run. The tremendous overheads—paying for the hire of all that complicated electronic and radio gear, for one thing, and expensive life-rafts as well as a boat: better fish detection and navigation gear, better quarters for everybody, more powerful engines to cut steaming time down all were very costly, to say nothing of the infinite variety of destructive and ever-rising taxation from which all suffered, skippers, mates, men, and owners—had not been nearly so high, even ten or twenty years earlier. British trawlers, too, had been excluded from traditional fishing grounds (where often they'd been the pioneers) by an excessive zeal for national protection. Because of the seemingly endless growth in working costs of all sorts and difficulties in distribution and marketing, there had been an increase in the size of ships and a decrease in their numbers. In Hull, distant-waters ships had declined from nearly 150 to less than 100 inside the past decade, so that fewer skippers were for ever trying to fill larger ships more quickly with more fish, and the road to skipperdom for mates was made more difficult because there were far fewer vacancies.

A hard life, indeed, in more ways than one, and for everyone. Fleets during that time had amalgamated and grown enormous and perhaps—in some cases—impersonal, in the process. Not Tom Boyd's, of course, or several others of the old school, born into and raised in fishing just as their skippers, mates, and men were. I'd met Tom Boyd of Boyd Line during the early days of the Second World War when he and I were in Coastal Forces—those not very satisfactory petrol-motor-driven craft which stopgapped as miniature frigates, gun-boats, minelayers, and every other kind of hazardous minor (and sometimes major) war-vessel through too much of the war. I never quite found out what I was supposed to be doing in this branch of the Service, but Tom was on the raid on St. Nazaire, a good, gutsy dash of great courage and endurance which succeeded because of its incredible audacity. Tom was there with a vessel called an ML, powered by engines requiring high-octane petrol stowed by the thousand gallons in large tanks, vulnerable and then quite unprotected.

For what he did that night Tom Boyd was awarded the D.S.O., a distinction almost without precedent for a lowly Sub-Lieutenant R.N.V.R. I knew he was from a trawler-owning family. His family away back had fished in smacks, on his mother's side out of Lowestoft, on his father's side ancestors included a grandfather who pioneered 'Klondyking' into the old Russia—shipping barrelled salt herrings, for which a huge market developed. Family tradition is that he was never paid for this. He died young. Tom's father had to begin as a butcher's boy but soon gravitated into fishing, with Thomas Hamlings of Hull. Business acumen, enterprise, a name for fair dealing and compassion for his men and their families—Hull owners founded their fishermen's widows and ophans' fund in the mid-1890's: the fund is still financed by both owners and men—gained him in time the managing directorship.

From this, Boyd Line branched out very successfully on its own. Its fleet today of ten sturdy big 'side-winders' all named with the prefix *Arctic* before words expressing something of the trawlerman's spirit—like *Crusader*, *Ranger*, *Buccaneer*, *Warrior*, *Corsair*, *Avenger*, and three large freezer stern-trawlers, the *Arctic Freebooter*, *Raider*, and *Privateer*—is a tremendous achievement for the old fishing family of East Anglian and Scots extraction. The same kind of courage, enterprise and endurance goes into this as Skippers and their men must know and call upon. Sleepless nights come not only at sea.

Skippers and crews at sea, after all, are usually harassed by the problems of the sea, which are often straightforward and clearly soluble, compared with the far greater number often unnecessarily made by the bureaucrats (and others) ashore.

Bernie Spicer was aware of all these things as he watched the gear shot with the usual efficiency, and settled then to drag again, with an alert eye on the ship, the sea, the other trawlers large and small, the lead of the warps, and the flickering faces of half-a-dozen instruments flashing or clicking away. One ear attuned to whatever of value might come over the loudspeaker set to the trawlermen's wave-length, the other to the noises of his ship as she ploughed along, all senses fully alert and occupied with the job in hand, he had no time to think then of long-range matters. But the days and the nights were long.

All the working trawlermen and the owners, managers, and everyone else in the industry were rather saddened at the time by what they considered as the appalling image of them sometimes presented to the British people by their public media, especially the all-embracing monster called T.V. The drama of the loss of the three trawlers *Ross Cleveland, St. Romanus* and *Kingston Peridot* had been a hey-day to the sensation-seekers and itinerant 'news'-mongers, a focus for inaccuracy, mis-statement and sometimes downright untruth. The real trawler-men were at sea, of course, their radios tuned as usual to their own professional waveband, no T.V. and no newspaper in sight.

"The real views of working trawlermen don't seem to reach the public at all," said Bernie quietly in a reflective moment. The Mate and the Bos'n nodded agreement. "We don't hold with the sensational stuff. It seems to us often slanted to make it sensational—the most dangerous sea life, the peril of 'black frost' and all that kind of stuff. Why keep shouting about that? It is part of the life—yes. But a small part we learned to live with years ago."

A lot of the stuff his people told him they'd seen and heard, afterwards, they knew came from the loquacious, found conveniently in pubs. "We could name some of them," he said. "They average one trip a year, if that."

When he was Mate of the trawler *Joseph Conrad* about a year earlier, he recalled, a team of five men came aboard for a winter trip to Iceland to make a film. One was very sea-sick but they all stayed on the job, though it was a time of vile weather and savage cold. They showed him the film afterwards. It was a splendid job—the *real* thing, ably photographed. But he'd never heard of it being shown.

Perhaps the moguls feared that its lively shots might make people seasick . . .

The cod-end was nice and full this haul. As it came up it floated, and the strong deck-lights picked out the fish-packed bag as it broke the surface with captured fish protruding from every opening in the big mesh—crimson 'soldiers', big fat cod, sloppy great 'cats', black halibut, a few haddock (scarce now on the distant grounds available internationally).

Such a bag must be brought over the rail and emptied on deck with skill and care lest the weight of fish causes damage by crushing. All hands, Bernie and all, concentrated their attention on this job, for they were determined to get those good fish into the fish-room in first-class condition, and keep them so. They filled a large pound and overflowed to pour into half another. Before the night was out Bernie was bringing aboard such hauls as flowed into all the pounds and so filled the cod-end that the bag had to be taken up in parts. The nets were designed for that. It wouldn't do to funnel hundreds of baskets, lifted high on the tackles, through the cod-end in one great finny torrent.

Good fish! We were on them at last, a week from the dockside. The lads, oilskinned, gloved, thigh-booted, waded in and began cleaning with a smile, happy to know that they would be cleaning, washing and stowing all night just with that bag, and hoping there were many more to come. Fill her up! She could never fill too fast.

There was the lugubrious voice of old Neville on the R/T moaning that his last haul was 50 baskets.

Bernie smiled.

Chapter 6

OFF BEAR ISLAND

THESE good fish didn't run for long. Maybe too many after them. There were soon a score of vessels busily trawling the area and, apart from a couple of smart little Norwegians, the *Arctic Vandal* was the smallest of the lot. We were out of sight of land and far from the tracks of all merchantmen, even those trading to and from Murmansk and Archangel and the Yenisei River. We were in a world of our own concentrated day and night on fish, and good marketable fish were no longer there in useful quantities at all. We were fortunate with the weather. It was not good, of course, but it seemed to be trying to co-operate. The new type of bright orange-coloured outer suits—all-embracing jacket, water-tight but flexible to work in, and bibbed trousers of the same light material secured with soul-and-body lashings around stout rubber sea-boots—kept the men dry in almost anything. This was the sort of suit which (with his own faith and courage) saved Mate Eddom's life.

[I wished it had been developed in the days of Cape Horn winter roundings under sail, for it would have saved untold grief down there—the one horrible sea-zone where good fish don't congregate. Even they have the sense to avoid it.]

If we weren't filling the pounds (or the fish-room) with good cod or anything else just then, at least I had plenty of time to yarn a bit with the lads, and to watch the fleet at work. Several stern-freezers were on the bank—fellows over 220 feet long, of up to 2000 tons, plodding along with their big trawls out directly astern. Only a few of the *Vandal*'s people had served in them, though they changed freely enough from ship to ship and all were well aware of the big fellows' advantages, and disadvantages. One of our spare hands, a cheerful tall Yorkie always ready with a smile, had been about rather more than the average trawlerman. He had tried the Merchant Service, too, mainly in the Union Castle Line, but he always came back to fishing in the side-winders.

"The money's good," he explained. "And you're sailing out of your own home port. Every time you do come to port, it's your own."

He was about six feet tall, strong-featured, dark, well built, and liberally tattooed. I called him 'Mild', as he had a couple of large beer pumps tattooed on his brawny chest, the one labelled 'Mild', the other 'Bitter'.

"That's all the beer we get," he grinned, admitting the design might not be approved among those tattoo artists the Japanese. I gathered there was at least one tattoo shop still operating in Hull, and it did good business.

"A tattoo is as good as a vaccination," declared Mild, though he didn't say against what. Mild was indeed well vaccinated. Across the backs of the lower joints of the four fingers of his left hand ran the word TRUE, on the right, LOVE. Another hopeful slogan crossed his toes. Skulls and cross-bones, an Indian head, stylised ships, clasped hands, anchors, ensigns, birds, etc., jostled for space up both hefty arms and on his shoulders. The other lads were mainly tattooed too, except the Skipper, though not as liberally as that earnest believer in frequent such vaccination, Mr. Mild.

But they didn't approve the art, quite regardless.

"I sat beside a young woman in a bus at Hull the other day, and her fore-arms were tattooed like a sailor's," said Radio Op. Norm Willis. "It looked stupid—pretty dreadful, I thought."

Perhaps the young lady believed in its good effects, too. She certainly was no fisher-girl. Only the Russians have women in their crews, and nobody had been near enough to them to see whether they were tattooed or not—probably not.

Mild had served several voyages in stern-freezers, but he didn't care so much for them. Too long at sea and too much like factories, he said. A fellow almost expected to clock in and out again, and go home by bus. Nor, according to his expert view, were they technically perfect yet. One ever-present danger—or at any rate, serious disability—aboard a trawler towing her large and costly nets along the floor of the sea is to have it snarl up on some wreck, still there in plenty on those Arctic grounds round the north of Norway and Siberia from the Second World War and its north Russian convoys.

It was Mild's view that the old-fashioned side-winder could unsnarl herself more easily than the stern trawler, mainly because, with the warps on one side, she could swing through 180° on that side, go back, and clear her trawl. The stern-hauling vessel runs far greater risk of getting the heavy wires fouled on her propeller. Her warps can cross, too, and otter-board trouble can be a headache.

So declared Mild, with the support of others with experience.

I hoped to find out more about stern trawlers for myself, and it looked as if I would get the chance. The skipper of the *Arctic Freebooter*, one of the Boyd big fellows, called the ship by R/T to say that he was going into Honnigsvaag, a small port just inside the North Cape where the Arctic trawlers sometimes touch when necessary. He could probably pick me up on the way in, he said, for the weather was tolerably good at the time and the state of the sea good enough for his outboard-driven rubber boat. But he was many hours away and the weather worsened quickly, as is its habit in those parts.

In the meantime Bernie, convinced that he was not going to catch a fill of fish in those parts on that voyage, picked up his gear after another wretched haul, stowed it aboard, rang full speed ahead and loped off towards Bear Island. We did not see the *Freebooter* then at all, though we might still see her on the other grounds.

Bernie had already been fishing then for four days. At most, he had another nine to get a paying cargo, and even the small *Vandal* had to make at least £6000 just to clear her costs each voyage. The pressure no skipper escapes while he is a skipper was on, and he had to get going. For my part, as the sea slapped the ship's sharp nose and sent protesting showers smashing over her and she bucked and rolled, I was happy enough not to be bouncing about in a rubber boat. I'd seen something of those things in the Hudson Bay Company's reconstructed bomb-ketch *Nonsuch* during the summer. In anything of a sea, they offer about as much protection as one of those large jellyfish seamen know as Portuguese men-o'-war, though unlike the jellyfish, at least they don't sting.

There were ice-jams and really bad weather not far away. Norm told me a large German motorship, not a fisherman but a freighter coming from the Kara Sea or the Obi River, was stuck in heavy ice up round Novaya Zemlya somewhere and had been calling urgently all day for an ice-breaker to get him out of it. There was a bit of a blizzard up there too.

"The Russians are sending the breaker," said Norm. "His chances are good."

I hoped they were, and was glad that the Hull distant watersmen—though unafraid to follow the fish anywhere—fished in waters where a touch of Gulf Stream Drift, even though well dispersed, managed to keep vicious ice out of sight for much of the year. It was different round Novaya Zemlya and the Kara Sea way. We bashed along in open sea. After a while, even a few stars shone.

The manner in which all hands in these ships calmly accept whatever may be the odds against them—and, indeed, never think of the risks of their calling—would perhaps astonish the landsman thrown suddenly among them. But seamen in fact see the works of the Lord, and know Him, just quietly and surely. They know and they accept and are not worried. It was the same in Cape Horn ships—those most glorious creations that yet were called killers sometimes, though not by their own people—in Antarctic whalers, in all ships in peace and war though war brings greater stresses and sometimes strange shipmates.

I gathered, talking with them quietly, that all the experienced men in the *Vandal* knew what it was to face sudden death. It was easy to go overboard from the pitching decks of an Arctic trawler, knee-deep in fish. 'Over the wall', they called it. The new deckie-learner had done that last voyage, some time in the long night watch, and nobody knew until the morning. The ship was steaming at the time on the way to the grounds. His was an unusual case. It was found that he had carefully packed all his gear to be landed, and, though a very young man, had written farewell letters to five young women. His seemed a tragedy of love, not of the sea.

Mild told me about one young spare hand who'd suddenly been flung 'over the wall', clung to a line until she rolled the other way, then climbed back over the rail with a big smile as if he had been enjoying the unusual view.

From off the North Cape to the grounds in sight of Bear Island is some 250 miles, a good 16 hours flat out for the *Arctic Vandal*. This was a long break for a small ship which could stay on the grounds at best less than a fortnight, once she began to fish. Bernie would have preferred to try other banks nearer, along towards the Murmansk coast and north of Kirkenes and Vardö in Norway. But the news from this general area was not good. The reason that he and all skippers kept their R/T turned up on the grounds was to gather news of fishing, and they were all experts at grasping the slight essence of fact filtering through the torrent of apparent chatter—not that it was all idle chatter, by any means. Code could be used: Bernie would suddenly prick up his ears at some innocuous remark.

"Why, the old so-and-so just hauled 500 baskets," he'd declare, though there'd been no mention of such a fact at all. "And he's off Bear Island."

Old So-and-So had a big stern-freezer. He wasn't fooling. Groups of the big fellows usually work more as squadrons than the side-winders do. If a group of those expensive couple-of-thousand-tonners, costing half to three-quarters of a million each, were finding fish to interest them round Bear Island, there should be some for us too.

There were. When we reached the new (to me) ground, we were the only old-style trawler there. Bernie shot, dragged, hauled—in plain words, put out his trawl, towed it along the bottom for three hours or so, then winched it up again—an average six times a 24-hour day. Sometimes the fish filled a couple of large pounds and spilled into a third: sometimes there were not so many. Rounded, sea-worn stones, rolled into the net by the bobbins, came up too and a few were kept as curios.

(Though they were commonplace enough.) At times we shifted ground a bit here and there around Bear Island, where the fishing charts showed the sea-bed cut liberally into pits and troughs where the fat fish lurked while they wolfed down sand-eels, until the slow thud of the trawler engines passed over them and the great open mouth of the trawl groped along, scooping them up.

Now the lads worked incredible hours. With fewer fish, 8 men of the 12 deck crew worked a 12-hour watch on while the other four had eight hours below, so that always four rested while eight worked. As pressure of work mounted, the system changed to nine at work for 18 hours while three rested for six. With a lot of fish to clean and wash and two men stowing in the fish-room, the other six sometimes couldn't clean the fish from one haul before the bulging cod-end was emptying more upon them. Each man picked up a fish as he stood among them in the pound and, with a few flicks of his sharp knife, had his fish half beheaded and skilfully disembowelled with amazing speed and dexterity. Then he tossed it (though it might weigh 20 or 30 lbs) to the washing trough, whence washing water and gravity in due course sent it down a chute below. He leant over for another fish and another and another the whole night on and the following day too, if necessary. There was no work bench, no team-work as I had seen for instance aboard the Portuguese codfishing schooners on the Newfoundland Banks and off Greenland. A man needed a strong back, an iron spirit, and all the guts in the world, or he quit fishing.

The bright working lights shone down upon them, the black sea beyond their range tossed fretfully without end. Up on the bridge, dark as the outer night except for the eerie light of the ghostly instruments, silent except for the sound of the wind and the blurp of the gyro-compass, there sits Bernie, alone, masterminding operations, watching everything. A fish-loop throws vivid green patterns like inverted palm-trees flashing swiftly on and off, on and off endlessly, the shape of the palm seen so briefly telling the watcher what is going on far below. In order that the skipper may watch both (and the Decca plot as well), nearby is another strange instrument, endlessly chasing a ball-point pen round in circles, the pattern of the circles telling the watching Bernie about fish. Yet another keeps an eye on the bottom (lest the wreck of the *Scharnhorst* should show up and rip all the gear—though in fact Bernie knows quite well where the battle-cruiser lies and avoids that area).

No one is at the wheel. No one is on the bridge at all, save Bernie, adjusting his body to the ship's motion, now and again taking a swift glance through the clear-view port forever whizzing round, glancing in at the radar plot and out for fishing lights, down again at the men working. In the little radio shack messages sing with their high-pitched whine and an Icelandic voice intones a weather forecast for all the seas up here as far as Spitzbergen. Our own fierce lights stab the foredeck's gloom: all else is blackness. From far below I hear the harsh sound of metallic repair where those stalwarts Ken

Skipper
(*Alan Villiers*)

Deckie learner
(*Alan Villiers*)

Bos'n Ted
(*Alan Villiers*)

Old Brian
(*Alan Villiers*)

and Trevor wrestle with some oily, cold steel thing turned recalcitrant but learning fast from them the error of its ways. (What watches they work down there, too! Those skilful, indefatigable, so often unnoticed toilers of the deep . . .)

So it goes through the long Arctic night. More fish, more fish! For the fishmongers and the merchants and the fish-and-chips shop patrons of all England, sleeping warmly in their quiet beds, while a little group of purposeful, courageous and almost tireless men work for their breakfasts—and, if they do well, a hard-won living for themselves.

Chapter 7

THE LIFE OF MILD

WE did well enough on the Bear Island grounds for the following week, not breaking any records and losing some hard-won cod through cod-end damage now and again—always put right with maximum speed and efficiency. There was no doubt that these trawlermen were magnificent workers. They worked the fish with rubber gloves but they went at the nets bare-handed. They never seemed to tire nor to lose their cheerful outlook even when fishing was poor, though their take-home pay depended on good catches.

"Oh well, we can't catch all the fish all the time," said Mild. "If we did we'd have too much tax to pay."

"I'm paying £10 a week tax now," said Eric, another single man.

I knew that spare hands like these in good ships earn an average £26 to £38 a week the year round. I'd seen wage sheets showing their average pay in the *Arctic Brigand* as just under £1900 for fifteen or sixteen trips a year (they all take a trip off for holidays). Some had tried shore jobs—mainly at the suggestion of wives, mothers-in-law, etc.—and hated them. They were paid much less ashore and had far more expense there, too. A trawlerman travelled with his work, not to it: he was well-fed and well housed (when he had time to sleep), had special medical care and an excellent organisation looking after him. The road to advancement was open, too: there was not an officer class reached by expensive apprenticeship or higher education. If he'd a head for maths at all and sound common sense—his sea sense could be taken for granted—it was up to him. The road to a skipper's licence was open, direct, and not particularly complicated (compared with a deep sea master's licence in the merchant service: a skipper was a deep sea master, too, and could—and many did—command big freezer-trawlers with no further licence). Steve, the *Vandal*'s deckie-learner, had an uncle in command of the big stern trawler *Orsino*, the vessel selected to act as guardian-ship for the Arctic winter fleets: and Steve intended to make the grade, too.

For my part, I wondered why the British government, which after all had been very slow in taking up the idea of using an assistance-weather-guardian ship for the trawlers at all, hadn't the sea sense to buy or build a real Arctic ship like one of the efficient polar-waters vessels of the Danish Lauritzen Line, which pioneered the type and in 1969 runs fourteen of them, from 1800 to 5000 tons. Vessels like Mr. Lauritzen's pioneering *Kista Dan* can go anywhere, but of course they are expensive to build and operate. There are no better working polar ships in the world: they could work the year round, in Arctic and Antarctic.

It seems a pity to take an expensive stern trawler away from her proper employment. After all, the governments of the great democracies do not seem to have developed much if any sea sense. Perhaps this is not a quality of much appeal to the mass of electors. Portugal and the U.S.S.R. do a great deal better in such matters. So do countries like Denmark and Norway. I have made Arctic voyages in two Portuguese fisheries mother-ships, the old and the new *Gil Eanes*, a motor-ship of 3000 tons fitted with a modern hospital, operating room, a met. centre, chapel, plenty of accommodation for rescued crews, and even stocks of frozen bait for the long-lining dorymen, of whom some 3000 still fish the Newfoundland and Greenland banks—all Portuguese, about a thousand of them still operating from big auxiliary schooners.

These Portuguese hospital-and-assistance ships were a boon to the dorymen. They cut down their loss rate tremendously, for both the old *Gil Eanes* in her day and the new *Gil Eanes* now conduct radar searches for lost dories. A senior officer aboard acts as a sort of admiral of the whole fleet of 36 or more dory-fishing schooners and motor-ships, not in any way controlling their fishing but looking after them, saving ships and lives—above all, helping them to fish better and to save themselves. Those Grand Banks and Greenland dorymen are terribly exposed. They fish often out of sight of their ships. Most have no power, and no radio in the dory. They must fill or at least part-fill the dory to have a worthwhile catch at all: but when full they can swamp. The men are heavily dressed against the cold and the biting winds. Often, at the end of a long day's hard fishing, they find themselves to leeward of their ship, unable to fight their way to wind'ard back. Sometimes a hundred or more used to be drowned in a night, Portuguese, Canadians, Gloucester-men, and Newfoundlanders alike, because their dories swamped and the sea beat them to death, or they died of exposure. The schooners then lay to big anchors with a long scope of cable: once fishing, they could not easily move. With their dorymen away, there was little or no crew left aboard to work the anchors.

The power-windlass helped greatly for then they could weigh and drift, clanging their great bells for the dorymen to hear and recognise. (Each ship's bell is distinctive.) Auxiliary power in the ships followed later, and this is a blessing, too, for then they could go to leeward and search. No longer must the dorymen undertake the murderous slog back to wind'ard. Marshalled by the assistance-ship, aided by her radar and radio—and by the determined skill of all the captains and the courage of their dorymen—losses were cut down and down until in 1969/70 it is unusual for the Portuguese to lose a doryman by drowning at all, though in 1969 eleven of their schooners and one last barquentine still made their six-month-long voyages to the Banks and many 1000-ton new motor-ships are still fitted only for dory-fishing. All the cod taken are preserved by salting aboard, and the ships stay out until they are full. This usually takes five or six months.

The *Vandal* lads knew these men. Some had met them

aboard their schooners sheltering from a hurricane at St. John's, Newfoundland, or fishing in their wooden one-man dories far out at sea, off Greenland.

Taciturn old Brian and the irrepressible Mild shared a great admiration for such men. So did the others who knew them.

"They fish a 12-hour day in an open boat about 14 feet long, flat-bottomed, with no protection at all," declared Mild, with wonder in his voice. "You see 'em away out of sight of their schooners. They put down long-lines and they have a bit of a little mast and sail to go back to the ship when they're full. They *stand* in their boats and jig,* when they're not working their long-line."

Mild told of passing among a small fleet of the dories on the bank called Store Hellefisk, off West Greenland.

"I shouted to one bloke to come alongside," he recalled. "It was good weather. He hadn't got much fish. We'd just made a champion haul, so I chucked a couple of quintals into his boat. I reckoned his need was greater than ours."

Whether the owner would have approved Mild didn't say, but it was a warm-hearted gesture. The doryman smiled.

Mild was in a stern-trawler at the time: the pressure of fishing against time is not so great in them.

There was no *Orsino* nor any other watching ship off Bear Island nor anywhere in European Arctic waters just

---

*'Jigging' was fishing with a double-barbed lead flicked swiftly up and down where fish congregated. The hooked barbs impaled them.

then. The month was September: the *Orsino* takes up her station in December. One of her main purposes is to warn trawlers—especially those operating round the north of Iceland in midwinter months, as the *Ross Cleveland*, the *Lorella* and the *Roderigo* had been—against the approach of dangerous icing conditions with on-shore gales.

At the time, the *Orsino* was going about her usual work, like four or five big fellows in company with us, among them the *Cassio*, *Lord Nelson*, *Prunella*, and the new *C. S. Forester*, on her maiden voyage. They crossed and re-crossed the banks, day and night, day and night, forever dragging, hauling, shooting, cleaning. I watched them from my vantage-point abaft the warm hotwater-boiler casing on the upper deck. Sometimes we passed within a few ships' lengths—always well clear.

One grey evening with heavy black squalls spattering the tumbling horizon, four of them were more or less close to us, each trawling on her own pattern, clear of all others. The nearest made a stirring picture, leaping and pitching into the black sea, the gulls flying round her by the thousand, the blaze on her deck lights fighting the stormy gloom. Not a human being was in sight anywhere, nor anything but the other eighteen ships by then assembled, the myriad gulls, the restless sea cold and hostile. The big ships and the side-winders went about their chosen ways apparently without human aid, almost as if they had grown there and been there forever, avoiding each other by some mysterious quality not contributed to nor controlled by man. Split-sterned, high-sided automated monsters and low side-winders dragging up fish, on and on they went, as if doomed to plough the sea up there for ever, or until they rusted apart and went down to keep even closer company with the fish they hunted.

"What chance had a fish up here?" I asked.

"Chance enough," Bernie said, a little gloomily. "We leave enough of them. Anyway, who ever heard of a fish that died of old age? They eat one another if we don't catch them."

He pointed to an excessively ugly monk-fish lying on the foredeck, a form of solidified sea-slime, tiger-teeth and ruthless cannibal habits.

"Little fish eat littler fish and big fish eat them. The whole of that monk's head is hinged on a body that's just a living fish-trap. His mouth is big enough at any time, but it can hinge back like a boa-constrictor's and swallow fish much bigger than itself. When you see them in the bag, keep your hand away! If he can't get the whole meal down in one gulp, he keeps the second bite until he can cope."

Bernie was right about the cannibalistic habits of fish, and right enough with the other view he sometimes expressed—that pessimists had been talking about over-fishing and fished-out banks and whole grounds for years and years, but men had been catching fish since long before the dawn of recorded history. There always had been fish and there probably always would be. You could go to a good bank many times and perhaps find no fish there, but then you'd come again maybe months, maybe years later, and there'd be the fish as thick as ever. Why? He didn't know. What the trawler had to do was find them.

The North Sea was supposed to be fished out years ago, said the Mate. It was true that some good fish might disappear from certain areas but plenty of nearer-waters and inshore fishermen still found fish. After all, there were 150,000 square miles of North Sea, practically all perfect fish habitat, rich in food for them, shallow enough to warm its surface a bit in the summer and deep enough to give the fish a place to live in the winter too. True, some of the best fish markets were within easy steaming distance of the North Sea banks, with access to 200 million consumers most of them fish-eaters.

"As for the fish, their whole life is to eat other fish and avoid being eaten themselves," he went on. "Look at those fat cod in the pound. They're fat because they've hogged down so many sand eels they can hardly swim. If they weren't here they'd soon be inside a big monk making it bigger."

Indeed the cod *were* bloated and they were full of sand eels. But over-fishing is a problem the working fisherman is content to leave to the scientists and the researchers—fortunately perhaps, an increasing body of highly competent men and women—and just get on with filling their own trawler full one voyage at a time.

Here and there, however, there is measurable evidence that fish are sometimes taken in excessive quantities. Some banks do become denuded, at least for a time, though it is not certain that this could be caused only by over-fishing. Those indefatigable world-wandering hunters not just of fish but of anything in the sea that is edible at all, the energetic Japanese, have discovered that even sound can be made a catching device. They have learned (or so I read in an international fishing journal) to make sounds like fish, by the use of a device called a Sonobuoy. First they picked up sub-surface fish sounds, sorted them out, then duplicated them and used them as 'bait'. Both whales and octopus, it seems, like the sound of pounding iron rods in water—an ear for 'pop' music, no doubt. Poor whales! As if they had not suffered enough from remorseless hunting already. They certainly can be decimated—big, friendly, sea-ambling beasts with defined habits and haunts and no defence at all. They can be hunted to the last big whale and, by 1970, quite possibly have been.

Demersal fish are different. For one whale, there is probably a billion of them, each capable of reproducing itself by the million. Cod, haddock, plaice, turbot spawn five to ten million eggs each year of their adult lives. They are all highly expert at camouflage, hiding themselves with perfect skill in whatever cover nature may provide. They are great hunters and hiders too: but they cannot hide very well from the ubiquitous trawl or escape from it either. Trawls have been developed that can be towed effectively at any depth where commercial fish may congregate, and others which can 'see' rocks or other damaging obstructions and jump the net over them—but still there are fish in plenty.

I watched the big, purposeful stern-towing giants, ploughing along in the sea with their distant silhouettes very like those of some modern war vessels, and the great open ramp cut into their steel sterns. Old-time sailors would have thought it odd to sail in ships with a runway for following seas built into them, but there have been such watertight ramps built into big ships since the Norwegian whale factory-ships started using them to fish the Ross Sea in the 1920's—the *C. A. Larsen* of Sandefjord with a bow ramp (not so good for whales, and obviously of no use whatever for fishing) and the *N. T. Nielsen-Alonso* with (as far as one knows), the first stern ramp, built watertight and permanent in her old tramp counter. This was a different proposition indeed. True pelagic factory whaling was then wholly practical to the ends of the earth, and the development of the ramp for trawling inevitably followed.

Most of the big fellows round us that night and many nights cleaned their catches in a simple 'tween-decks plant, freezing them at once in blocks of eight or so and stowing these down below until they had a cargo of 600 or 800 tons which they then carry back to port. These are the freezer-trawlers. Others carried the process further, filleting and skinning the fish and quick-freezing them in packs ready for marketing. These are the factory-trawlers, big ships like the *Coriolanus*, fitted with an expensive and elaborate factory deck where the gutting, cleaning, skinning, and filleting are all done by machine. Still others, like the new *C. S. Forester*, are part freezer and part wet-fish (i.e. fish preserved for a limited period in loose ice, as in the *Arctic Vandal* and other side-winders). These operate the same way, with the stern trawl and the 'tween-decks working plant, but save some capital investment and cater for both the deep-freeze and wet-fish markets by filling with frozen fish to capacity first, then spending the last ten days or so of each voyage putting the catches made then in loose ice. Unlike the older wet-fish types, they can make more ice, but preservation by this method cannot be too greatly prolonged.

These modern types waste nothing. Like all trawlers today, they carry efficient cod-liver oil plants, and many have fish-meal plants as well to convert bones, skin and other offal into food for broiler chickens, which thrive on it.

With ships like these, the ancient business of fishing has come a long, long way since cities like Tyre, Byzantium, Venice and Amsterdam were founded upon it, and, some 2,000 years later, Britain's industrial revolution so greatly increased demand and markets.

It seemed almost incredible that in the highly modern fleet of 19 ships round us up there, worth between them at least ten million pounds, manned by some 500 men, there were still a few old hands who had been brought up in smacks out of Grimsby, Fleetwood, or Lowestoft, fishing the Dogger Bank.

Chapter 8

OLD BRIAN

Like our own Old Brian—not that any of his shipmates called him that. Brian Taylor admits to 62—"Another couple of years and I've got my first half-century in fishing," he says—but he is as lively as any of them, a sturdy, stocky sea dog in his thick jersey and fisherman's smock with a knitted woollen cap jauntily atop his grey head. Grey it certainly is, with good cause, but the hair is still all there, and the blue eyes that for so long have looked kindly upon an often hostile world are still kindly, too.

Aye, Brian began in the smacks. A smack used to be a single-masted sailing fishing-vessel, deep-keeled to grip the turbulent North Sea waters, superbly buoyant and beautiful of line in her hard-worked hull the surer to live in them, with sail-plan manageable but ample to hold her off any lee shore the Lord did not intend she should be washed on. Later they were larger, up to 90 feet long and more—two-masted, ketch-rigged, with a bit of a donkey-boiler and winch to help handle the heavy beam trawl. They were superb little working vessels—tough, rough, and handy: and there were so many of them before steam took over what they were accustomed to work in large fleets, under a hoary old sea-dog of an outstanding smacksman known as the admiral.

"He *was* an admiral, too," Brian told me. "Everybody did what he said. He signalled with a big red flag and everybody had to know those signals and jump to 'em. They just couldn't drag beam trawls around under sail any old way, some shooting and some hauling, some on this tack, and some the other. They shot and hauled together and right smart too, on the proper signals, and the admiral said how they'd sail. It must have been a sight with hundreds of them on the job, all industrious fishing. I never saw hundreds myself. I was a bit late for that," he added regretfully.

It was just after the end of World War I, in 1919, that Brian began fishing, aged then barely 14. A lad could begin in steam trawlers then of course, if he were soppy enough (Brian said) but his father and all the other older fishermen advised him to begin the hard way—the *real* way, they called it, under sail.

"Learn the real way first, they said, as if these new-fangled steam jobs mightn't last. Tho' that wasn't the reason. Reason was a boy learned *proper* in a smack, about the sea and about fishing and about looking after himself, too. There'd been steam in trawling for 50 years when I started, ever since a couple of old Geordie paddle-tugs took to towing a trawl themselves instead of just towing smacks with their trawls out in a calm. That was around the 1870's. First they towed the smacks in calms and the smacks fished with their own trawls. Then some smart-alec figured he'd just as easy rig a trawl from his tug and do the fishing himself. So he did that—made money, too. They'd lost a lot of their towing jobs when steamers took over so much of North Sea and deep sea sailing-ships' trade. There was a lot of these paddle-tugs around, built of wood.* They never ought to've been allowed to do it, my old man said. They'll spoil the grounds and they'll spoil the men, too. And they'll be bringing a lot of half-frozen stuff to market that won't taste like fish at all."

Like all the old-time smacksmen, Brian had to serve a four-years apprenticeship before he could call himself a fisherman.

"I was cook first," he told me. "Boys was always cooks, but they was deckboys too, and working hard at everything in a year or two."

As for the cooking, I gathered that was arduous but not especially skilful except in the matter of keeping pots and pans on the small stove in storms. The crews ate well but rough—plenty of fried fish, of course, and a bit of roast beef on Sundays so long as the beef lasted. The fresh might last two Sundays. After that it was salt, and that could not be roasted. It was always boiled. There might be a ham or two if fishing the previous trip had been good. There were good stocks of ship's biscuit—

---

*Brian Taylor had his history right. In 1877 William Purdy, skipper and part-owner of the paddle-tug *Messenger*, an old-timer built in 1843 with a 25 h.p. engine, converted her to the simple trawling of those days by buying a set of trawl nets from Grimsby and a suitable old spar for a beam, and had a pair of trawl-heads fixed up for him by a local blacksmith. His investment in this lot was about £19 10s. (according to Edgar J. March's classic *Sailing Trawlers*). All the die-hard smacksmen jeered at the idea, of course, but the *Messenger* paid from her first trip. She had only a crew of three men, and no shore 'overheads'.

hard, tooth-breaking stuff–cheese, dried fruits and carraway seeds, a score or so cabbages to start with (no matter how long the trip was: it could be eight or nine weeks) and there were always plenty of strong condiments to help the salt beef down. Nobody taught the boy-cooks anything before they began that difficult job in a tossing, cold, wet and (at times) verminous smack bouncing about with a crew of hungry mariners of the most conservative eating habits on sea or earth.

"Only thing they ever showed me how to make was tea," said Brian. "That was easy enough. All I had to do was keep a kettle on the boil just about always, and whenever the men shouted for a brew, chuck another handful of tea into it. Once I took some old leaves out. I got a cuff then. 'Never take 'em out,' the skipper told me. 'Put more in! Don't waste good tea.'"

It must have been an unholy brew after a week or two, though Brian said he did ladle out a spoonful of mashed elderly tea leaves now and again, by night when no one was looking. Even then, it was only when he had to, to get more in. He learned how to make a sort of unleavened bread, too–a kind of Australian damper. All this had to be was hot. Washed down with smacksmen's tea, nobody noticed the fish taste in it or ashes, or anything else. They were hungry and it filled them.

The men treated him kindly enough, though it wasn't always like that for boys. Some were cousins or uncles.

"There was a good spirit about in them days," said the old man. "We was there to catch fish and it was hard work. A man had to be a sailor, too. You could take an

M.V. *Strathyre* I.N.S., later totally wrecked (*"Northern Scott"*, *Elgin*)

Scottish drifters leaving harbour in the morning mist (*National Maritime Museum*)

interest in that. You had a few easy times. If it was calm you couldn't drag the trawl along because it'd stay in the same place, with the fish swimming in and out of it. So we wouldn't put it down until a bit of wind came. Maybe you'd long-line* then, or jig up fat cod if they was plenty about. Manhandling the big wooden beam that kept the trawl open and handling all that gear was a lot of work, even though we had a steam-capstan to help. It was easier long-lining. You bait the hooks on all the snoods, lay the line out proper, and wait for the fish.

"I had to learn about all kinds of fish and fishing, and sails and looking after 'em, and steering by-the-wind and all that, and box the compass and recognise all the lights an' that and the headlands and places to get shelter in storms (if we could), and pick up all I could about the banks—where the fish was, how you'd know, and about the clouds and the look of the sky—what was going on up there and what it was going to mean to us down below.

"You had to train your eyes to *see* it all—not just cock half-an-ear to listen to some feller yapping on the BBC that had never learnt it anyway. It's all different now. Then we was relations, like them sensible fellers up in N.E. Scotland that follow the fishing still are. Now it's Tom, Dick, and Harry, an' as soon's they've left the dock half on 'em want to be back again. We was *fishermen* then, and glad to be.

---

*Long-lining was (and is) fishing with a stout line laid along the bottom with baited hooks at regular intervals attached to short pieces of lighter line called 'snoods'. The line was anchored one end, buoyed the other, or kept aboard. It had to be put down skilfully, according to the movement of the water and the run of the fish.

"Now you're 'old' at 30. You're only just starting to be an experienced hand then. You're beginning to get an idea what you don't know and you're a bit of use at what you do. You're careful, and you can do your work. That's the way it was, anyway."

That was the way it still is, too: I could see that, but I liked yarning with Brian. He served his full four years, then was rated spare hand. He began in the Fleetwood smacks, later moved across England to Grimsby. The Fleetwood smacks fished the Minches, the north Irish Sea, and around N. Scotland. Some had wells–large compartments in the hold open to the sea but with no openings large enough for fish to escape through–and kept their fish fresh. When they made longer voyages such as up to the Faeroes, they split the cod and salted them, for people still had the taste for good salt cod then (just as some inlanders still knew what really fresh fish tasted like, for live fish in wells travelled by special water waggons to inland markets by train) and there was a good market for it.

"We'd bring fresh fish as far as Harwich," Brian remembered. "We often wouldn't get home for eight or ten weeks at a stretch, and then only to reprovision and be off again. We looked after the smack in port too then. She wasn't a machine you jumped off in the locks, and left to somebody else. We always tried to fish the year around–we had to, to hope to make a living. It was plain common sense to keep the smack and her sails and gear in the best order we could. We was all brought up to do that."

Brian still had the habit, but the fully mechanised *Vandal*—her gear all wire, her nets all artificial fibre—had little need of his old-style sailoring skill. Like all the other British distant-water trawlers, she hadn't even a small mizzen to help her to lie-to. I wondered sometimes whether such a steadying sail might not have saved the *Ross Cleveland*. It would have helped to keep her head to the sea.

The world Brian Taylor remembered had been a different world for everybody, simpler, not so confusing for a man. An off-shore or inshore fisherman was a craftsman, and content at that. Brian said they couldn't get a bath aboard a smack at sea, or even a wash except maybe in rain during a calm. Living conditions were thoroughly primitive.

"Ashore, we used to wear a good guernsey and a white muffler on Sunday, strong serge trousers so thick they wouldn't take a press (they didn't get the chance), strong lace-up boots, a decent cap," said Brian. "That was our shore rig. Our working rig was the same but older—maybe patched quite a lot—with an oilskin smock, and leather sea-boots that lasted years, and a sou'-wester. That was all we had, but it was well made. We oiled our calico oilskins ourselves, and we looked after the boots. We didn't need to keep buying new ones.

"It usen't to cost us much to live. Funny thing, I don't remember ever hearing of any bad icing up, stability problems and all that then. We didn't go to the Arctic, of course. The Fleetwood and Grimsby smacks were very seaworthy, and they sailed well. Same with Lowestoft jobs too, and those others from down-Channel—out of Brix'am and the like. They had a right proper sea form that'd live in the sea—under God, of course. They was strong built an' deep, an' weatherly. Any ship'll get lost if her number's up. There was a place out there where good fishing was, by the Dogger, that old smacksmen still called the cemetery when I was with 'em. So many of them are there—sometimes hundreds after one storm. But it had to be a rip-snorter, with the wind setting up a shocking mill-race across the tide. Only way to live then was get out of it—fast. To be caught in it at all could be too late."

Indeed it could. When Brian first went to sea there was still a strong memory among the smacksmen (and their brothers in the powered fishermen too) of a night on the Dogger in the 1880's when 360 of their fellow seamen died in one savage, long-continued gale, for even the stoutest smack caught there could be quickly overwhelmed. With her sails blown away, her decks swept by the sea, companion-ways stove-in allowing the sea to smash into her hold, she hadn't a chance. Tough old Brian's tough old countenance with the profile of a fighter softened a little when he thought of all the dead fishermen lying there so long, hostages lost in the sea.

That North Sea can be a savage place for those who seek their living from it: its best fishing banks can throw up the most viciously dangerous seas.

"I never saw a snarlier sea than around that Cemetery in a cross-tides gale," said the old man.

In 1969, fishing craft were still lost there. The Scots

seine netter *Coral Isle* went with all hands that December.

Smacks fished the year round, too, just as powered trawlers do, but over the worst North Sea months—from just before Christmas until after the March equinox—many moved inshore the easier to run for shelter when they saw a bad storm coming out of the east. For the gale then could blow them ashore. They did not all use the beam trawl: some old-timers stuck to methods going back at least to the days of St. Peter.

One most dangerous aspect of the old-time smacksman's life Brian had been spared. That was the dangerous boat-work essential to transfer boxed catches to carrier cutters or steamships, in the open sea. The fisherman's meagre living depended on the catch and getting it to market. It was uneconomical for the big fleets to go back every week or so, or they could never hope to pay at all. So fast carriers were hired to come to them and rush their fish back to market, the nearer to Billingsgate the better.

Scores of smacks could not very well come alongside the carrier and toss their boxes aboard, for they would jump about so much—the carrier rolling one way, half-a-dozen assorted smacks the other—that dismasting and structural damage would be inevitable. Every smack had a bit of a boat and her men were good oarsmen, and the practice was that they all launched their boats and pulled for the carrier as fast as they could, eager to get their fish off at its best. Many, many men were drowned in this wild boating, good boats and superb seamen as they were—in the aggregate, quite likely as many during an average year as were lost from foundered and wrecked smacks altogether. The heavily dressed fishermen stood no chance in the sea, either way.

The admiral could stop the fish transfer by signal, of course. His word was law. But he was very reluctant to do such a thing. Too many livelihoods depended on getting those fish to market. The fisherman has always been an optimist—a bit of a fatalist, too. The outlook goes with his calling. So he will take his chance and not remember how vulnerable he can so quickly be—until he must.

By the time I had learned what I could of life in the days of the smacks, the *Vandal* was homeward-bound. She had fished for a week or so off Bear Island with varying luck. Then she moved down to the coast of North Norway, well off-shore, to try some of the banks there on the way home. Picking up a few hundred more baskets of good fish would be a help, for one voyage of such a ship cost in 1969 more than the capital value of half-a-dozen smacks. They could scrape up a few fish and live. The *Vandal* had to take them by the thousand kit (a 10-stone standard measure) and fast.

She had fifteen or twenty years to earn not just her own capital costs but money enough for her replacement, and shipbuilding costs climb and climb. Good smacks could last for 50 years. In 1969, a few tough old Grimsby smacks of Brian's early days still fished from the Faeroes.

Now our fishing was done. It had not been a hard trip. Indeed, as far as their pay was concerned, all hands could have put up with much more arduous conditions. She had about 1,300 kit of good fish down below where there was room for over 2,000. But they were as cheerful as ever—Mild, Eric, Brian Clark the Mate, Ted Newman the Bos'n, Ken and Trevor the engineers, Norm the Radio Op., Steve the Deckie-learner, Pete the galley-boy. The fishing was done, the gear cleared up and made ready for the next voyage, the steel bottom-rolling bobbins changed, the whole net, wings, square, belly, cod-end and all gone over thoroughly. These steel bobbins take a beating from the Arctic bottom, and sometimes wear thin on just one voyage. The ship was cleaned up thoroughly fore and aft, working deck, bridge, quarters—everything. Fire and boat drill were carried out energetically.

With a full ship homeward-bound, good morale would go for granted: but this time no one's pay-off would be very large. The wage families had drawn during their absence would very likely be all they would have. Yet the whole crew was as cheerful as ever. With the fore-deck clear, the gear stowed away ready for next trip, her discharging berth allocated in the fish dock, the *Arctic Vandal* hurried along under grey skies towards the mouth of the Humber. When all else was done, the lads played a kind of football with a worn-out bobbin on the fore-deck, old Brian along with them.

I knew that he must be one of the last working trawler-men from those now far-off days, but the tradition that those hard-working, hard-sailing, simple men had so thoroughly established—the cheerful acceptance of life as they find it come what may—goes firmly on. Life as Brian and his contemporaries once knew it has changed indeed, in many ways greatly for the better. But the deep-rooted spirit of cheerful, unworried optimism lives firmly on as the often lovely names of the old smacks still do in nearer-water fishermen—the *Star of Hope*, *Happy Return*, *Ocean Gleaner*, *Bright Hope*, and *Tranquillity* —thought-over names, pregnant with meaning.

In the cheerful freemasonry of the seagoing fishermen's lives, so much that is admirable is taken simply for granted. A man has his shipmates and his ship: if any worries about the high risks of his calling it is not he, though he is well aware of them.

Chapter 9

STORY OF A MISSION

In one way, the old smacksmen were better off than their descendants in the fast, expensive trawlers of today. They had Mission hospital ships out on the grounds with them, at first little ships such as they sailed in themselves but big enough to carry a good dressing station, some beds for the badly injured and ill, and a friendly first-aid trained missionary in charge of the lot, at their service. Such hospital-assistance ships first sailed the stormy North Sea waters in the early 1880's. At first rather primitive themselves, they did their best. Their well-stocked first-aid chests soon grew to cabinets, and then to cabins. By the early '90's, surgeons were aboard. At the turn of the century, there were eleven tough little assistance-ships out there with the fishermen, four of them with medical staff aboard and fullest practicable facilities.

These did splendid work for twenty years. When steam took over from sail, a steam hospital-ship was built to keep up with the new-style power-driven fleets (many earlier steam-trawlers continued the old practice of several of the bigger ports of fishing in fleets, under an admiral). The 275-ton *Alpha* was built of steel at Leith to serve then, in 1900, at a cost of £12,000, and she was soon followed by two others, the *Queen Alexandra* and the *Joseph and Sarah Miles*. There were then four principal fleets, and the plan was to have a hospital-ship attending on each of them. There was no R/T then to call them when needed, but each fleet worked in sight of its own admiral, and the Mission ship stayed with them.

This was no government enterprise. The ships were financed, built, manned, and operated by the Royal National Mission to Deep Sea Fishermen. The story of the Mission—a go-ahead, down-to-earth, energetic and effective organisation which seeing a need both urgent and unmet, met it, and goes on meeting it—is stimulating.

It all began with a visit to the 'fleeters' smacks on the Dogger Bank in 1881 by a Londoner named Mather, who until then had had nothing to do with fishing at all. Ebenezer J. Mather was a worker in the Thames Church Mission. Seamen he knew but fishermen he did not, for the great number of these—12,000 worked in the North Sea alone—did not come up the Thames. Carriers brought their fish to the great fish market of Billingsgate. By this means it was simple enough to get out to visit the fishing fleets. Mather went in the steamer *Supply*, carrier to the 'Short Blue' fleet of East Anglia. She took out 2,000 large empty boxes, lots of suitably broken up ice (then usually imported in the cut block or by barrel from Norway) and brought them back full as fast as she could go. The 'Short Blues' he found were over 200 fine smacks of 50 to 80 tons, a beautiful sight with their tanned sails turned scarlet in the setting sun. From these came at once '400 of the wildest fellows I have ever seen,' two to a little boat, pulling pell-mell to be first alongside the *Supply* to collect boxes for the transfer of their fish. Small herself, the carrier dipped and rolled in the restless sea without way, and the boats leapt on the tops of swells or sank deep in the troughs beside her in grave and constant risk of fatal collision, while the whiskered wild men in them worked their oars and shouted as they manoeuvred for pride of place to get alongside. Heavy boxes tumbled among them. A tumult of oaths arose, at full lung-power. Mr. Mather watched and listened with amazement.

The brief trip did far more than just change his life. The more he learned of the conditions under which this great storm-tossed fleet of little ships and its crews then had to work—cut off from their homes and the land, in peril in every sudden squall and sometimes decimated in winter gales, without aid of any kind save such as they might be able to provide for one another (and always ready with that), with no access to medical care except briefly at times in the waterfronts of smack-jammed ports, hard-working, hard-living, hard-swearing too when they felt like it, it was obvious that these were a great band of neglected seafarers working under God but not particularly then in the care or thought of any of God's servants.

Once on the fishing grounds, too far from suitable havens to run for shelter, they must fight it out or founder, get their families' livings somehow or see them hunger.

Conditions seemed really intolerable. Curse number one were the foreign carrion-craft called 'copers', coping with and bringing nothing but evil—the floating grog-shops of the cold North Sea, forever ready to supply fiery spirits at duty-free rates which though low gave them high profits. In calms and light winds when the smacks could not fish, the copers were among them, and the smacksmen's boats made for them to buy a little temporary forgetfulness; a bleary, seeming relief from the harshness and apparent hopelessness of what then were basis of too many of their lives. This drunkenness could ruin good ships and fine men, and often did. The 'copers' sold cheap tobacco too, and were the only source of the solacing weed. Many a smack's crew pulled over to buy a bit of 'baccy with their hard-earned pence, to stay and pour down raw Hollands gin until they were stupid. Money all gone, the 'prentice-boy would have to pull back to the smack to bring across their bit of gear for barter—anything and everything went, even at times the smack herself, for the gin was maddening stuff and the men had enough for temporary forgetfulness but never to get used to it. The 'coper' would move on to other prey, present in such careless abundance. Men were drowned in drink, smacks lost.

Mather heard of this, and saw something of it. He realised it was a great, mad tragedy of neglect. When that carrier got back to Billingsgate he knew what he had to do, and he did it. He founded and organised a mission for fishermen—the Mission to Deep Sea Fishermen, soon to become both Royal (Queen Victoria was strongly for it) and National—to go out among the fishermen at sea, and *stay* there with them serving them in body and in spirit. Within a few months, the Mission's first assistance-smack sailed out to sea—the little *Ensign*, bought and roughly converted for a subscriber's £1,000. Her small band of people included a skilled medical dresser and a minister, and she moved among the smacksmen offering a better and infinitely more useful diversion for them than any 'coper' did—far more than that! The fishermen were delighted. *Somebody* cared.

But some continued their patronage of the floating saloon, saying it was the cheap tobacco they must have—true enough, but the liquor swiftly followed. British Customs regulations forbade the Mission ship to carry duty-free tobacco for the fishermen. To do this, she must make a voyage: all right, so she would come out to sea and make a voyage first to a Belgian or Dutch port where she picked up tobacco so reasonably the Mission could undersell the 'copers'—and did. Tough big Victorian smacksmen, bewhiskered and begrimed, swiftly came for their 'baccy and care for their cuts, wounds, and rheumatics, stayed to ask for services. The missioners brought Bibles, reading matter, warm clothing, but neither sectarianism nor bigotry. They carried the message of Christ in simple words: and these the fishermen could understand. They succeeded from the start, though a few foolish jeers had followed the *Ensign* and her men when they first moved out to sea from Yarmouth, mission flag bravely flying. Soon learning of the good they did, no one jeered again. Within a very few years of

Founder Mather's trip in that Billingsgate carrier, the drink-peddling 'coper' was not just out of business but driven off the seas for good by international convention. It was a resounding victory, a tremendous achievement for Ebenezer Mather, the quiet mission worker working for Christ.

Among the early medico-missionaries in the stout little Mission ships was one the world came to know—Dr. Wilfred Grenfell, the famous Grenfell of Labrador. Dr. Grenfell was indeed the first actual surgeon to work regularly in a mission ship on the North Sea grounds. Back in 1887 the Mission already had seven of its own vessels, financed to a considerable extent by mortgages to the smack-owners, who were well aware of the value of the work they did. They also helped to clear expenses by fishing like other smacks, on suitable week-days. Medical care was still out of a copious first-aid cabinet or cabin and the skill of specially trained skippers. As soon as possible, the mission set up a better medical department. Dr. Treves—later Sir Frederick—surgeon at the London Hospital (which among others treated fishermen casualties brought to Billingsgate by the carriers) came in as chairman of the new department. The young Grenfell, well known to Treves, was just the man to go out as the first smack-borne surgeon when the Mission smack *Clulow* was fitted out for an experimental two-month period. Energetic, enterprising, an experienced yachtsman in his own right, well used to fishermen, to living in little ships at sea, and to getting on with tough chaps found there (or ashore: he'd already run the first East End boys' club in

London), Wilfred Grenfell was the perfect choice. Mission funds were not great and obviously, as a hospital ship, the *Clulow* could not help her costs by fishing. But the work being done caught the public imagination in a way the fishermen had never done before, despite the appalling calamities which now and again had overtaken them. The widows in their despairing black and the helpless orphans were seen only by the folk of East Coast ports, who were too used to them.

The *Clulow*'s first two months under Grenfell were a great success, despite the fact that at first he found the life hard and keeping the sea bouncing about in a smack on station in January very different from summer yachting. The fishermen crew found their surgeon an odd shipmate too, at first. He took cold baths in sea water on the foredeck every morning, exercised himself running up and down the rigging, and was always up to something. They had never met a man like this before. The sea gave them cold baths enough, and their smacks sufficient exercise. But they liked him rather as the Bedouin liked T. E. Lawrence, with something of a sense of admiring and astonished awe, but soon deeper than that. For Grenfell brought Christ among them simply and clearly too, and this was a profound experience.

Surgeon Grenfell found much to do, and did it. News of the good work being done out there on the desolate North Sea spread slowly, but it spread. Funds came in for the *Clulow* and the Mission's work. Grenfell became superintendent, moving ashore. More little hospital-ships were commissioned. As he was a new kind of evangelistic gentleman come among them at sea, Grenfell was a new sort of man in their home ports, too. He fought the evils he found there, and of these the first was drink. He set up the first Mission hostel in an old seamen's hall at Gorleston. As an evangelist, an organiser, a champion for the fishermen, Wilfred Grenfell was unique. He was also hard to hold: the 150,000 square miles of the North Sea and all its adjacent coasts could not suffice for his abundant energies and imagination. The Mission was already, by God's grace, successful there.

Within a few years, by 1892, Grenfell had moved on to Newfoundland and thence to Labrador, sent there first among the fishermen by Founder Mather's Mission. There had been many trans-Atlantic requests for help from there: he began at once to fling all his gifts into the performance of great works there and laying the foundation for more. So bright a star must shine beyond the confines of the cold North Sea: but the light the Mission first brought there shone steadily on. When Grenfell moved across the Western Ocean, the Mission to Deep Sea Fishermen's hospital-ship service was well established, and so it continued for many years until not only the fishermen wondered how they had ever managed without it. When steam-powered trawlers came, the hospital ships were modernised with them: when new and more distant grounds were discovered and worked (some of them a thousand miles beyond the North Sea), the little mission ships did their best to chug up there too.

But with ships fishing singly, no longer in marshalled fleets, and their capital and running costs steadily rising,

Old-timers at Grimsby some years ago *(Alan Villiers)*

Admiral Foot of the Gamecock Fleet

the private provision of an effective seagoing assistance service for the distant-waters fisherman became more costly and difficult and, at last, quite impractical. By 1910, the Mission was coping with a recorded total of some 15,000 cases—surgical and medical—at sea each year, which seems incredible. Many were minor, of course—perhaps the great majority—but if the Mission's little *Queen Victoria* (she had a ten-bed hospital), *Albert*, *Clulow*, *Alice Fisher* and the others were not there, a considerable number of these could well have become major. The smacksmen (and their successors in powered trawlers) believed in the oddest of remedies. For them, a pill of Stockholm tar and flour rolled in a husky fisherman's fishy hands was standard treatment for most ills. 'Ointment' of coarse soft-soap mixed with sugar was applied externally on anything. A swig of plain turps was thought good for the kidneys; at least it gave the taker something else to think about. It was the seagoing fisherman's way to work on with anything, infected fingers or ghastly sea-boils (caused round wrists chafed by coarse oilskins constantly worn), broken ribs. The smack had no place for the too readily incapacitated.

The Mission ships held simple services too, and the smacksmen crowded to these in fine weather. They presented a little music, distributed books, warm clothing, tobacco (at cost): and they provided that rarity in the smacksman's life—a thoroughly clean ship fore-and-aft, below and aloft, manned by men who were able to wash in fresh water and change at least some of their clothing now and again.

But the 20th century had little place for fresh fish brought by sailing-smack. The picturesque craft lasted a long time, but their old ways were changed–no more 'fleeting', no more carriers, no more tremendous assemblages of tan-sailed little ships tossing on the sea. Such smacks as survived fished locally, or were loners, 'single-boaters', on their own, using salt or a well to preserve fish. By 1915, 600 trawlers worked out of Grimsby alone. Powered trawlers moved farther and farther afield, were more and more difficult to serve even by ships of their own type. A good smack cost a few hundred, but a full-powered trawler cost many thousands. How many would be needed to offer full, year-round, seagoing service to a thousand busy, hurrying, widely scattered trawlers, working independently? Trawlers sailed from ports scattered round all England, Scotland, Ireland, and Wales, from Fleetwood and Milford Haven, Lerwick in the Shetlands, Stornoway in Lewis, Aberdeen, Granton, Lowestoft, Llanelly, Lyme, Wick, Whitby, Wisbech–from a hundred ports small and large, to fish a hundred banks. How could a mission serve all these effectively, the year round, from little ships at sea? There was no answer: an assistance ship for later days must be big (like Portugal's *Gil Eanes*), provided and manned by government agency.

There was still plenty of work for the Mission to do. The modern trawlers were more often in port than the old smacks were: they had to be, to land their fish while marketable, to restock with ice and coal, and stores. The sick and the injured need not be kept aboard too long.

Ashore, too, the trawlerman needed help—a need which included the wives, widows, children, the retired, the 'scrapped' (a trawlerman grew old fast: he still does), the disabled and the fit alike over the few days of their liberty.

Early in 1899, an anonymous friend in Scotland gave the money to put up a Mission hut by the Grimsby Lock-pits, where the trawlers locked in and out in the tide. Here the service was among all the fishermen who passed that way, inwards and outwards. Grimsby was—and still is—the largest fishing-port in the U.K. There was a big movement of men and lads coming and going, wide-eyed youths to find work in the industry, cheerful fishermen homeward-bound with good pay-packets. There was no shortage of those ready to batten on them: the cheerful trawlermen themselves, rugged and tough as they were, could be easy prey. All knew the Mission by this stage. Competent and sympathetic missioners who had been at sea and knew these men could do excellent work among them—and did, and still do.

In Mather's day it was obvious that the *urgent* need was for serving ships among the smacks at sea and this was met magnificently. Today's high-powered distant-waters trawlers, in constant radio touch with the land, can and often do achieve quiet wonders with urgent accidents or sickness cases by expert advice quickly sought and quickly given over the air. Owners operate medical centres near the lockpits at the larger ports. There are checks and examinations; there is the National Health Service. There are specially appointed medical agents in convenient ports. Nonetheless it is to be hoped that a real fisheries assistance vessel fit for the Arctic winter, designed and fully equipped to furnish up-to-the-minute weather information, to offer rescue facilities, and to cope with serious accidents where and when they occur, may be nationally provided before too long. The three little ships provided in the 1880's by the R.N.M.D.S.F. cost between them, with all their equipment, some £30,000, and annual running costs ran at round £1,000 for each ship: a budget for just one suitable ship today would necessarily be enormous. She could not be built for *half* the cost of all the mission's busy shore establishments.

From that Grimsby lockpits Mission Hut the shore establishments spread, and spread. Today, 70 years later, from Lerwick in the Shetland Islands right round both coasts to Newlyn by Penzance in Cornwall, the Royal National Mission to Deep Sea Fishermen maintains fine modern buildings, each a combination of hostel, recreation centre, up-to-date hotel, quiet chapel, and reading-room—centres for social gatherings where the fisherman old or young has always a pleasant place to go and use and a friendly Missioner will look after him according to his need, and wish. And be there to help his family, too, in case of need.

At Grimsby today, within sight from those same old lockpits where the heavy-laden little trawlers slip quietly out to the wide brown river and hurry off to sea, or lock in rusted and sea-stained with weary crews, there stands a splendid new building beside the old street called Hope which must be one of the most modern of seamen's mission buildings in the world. The only evidence of its purpose is a great picture of a trawler worked in coloured stone high on the front wall. The cheerful cloth-capped, felt-hatted or bare-headed men and youths passing through the door have no obvious mark of their calling about them, as Brian Taylor's generation and predecessors had in their jerseys and general rig. They may include Icelanders, Faeroese, Norwegians, Danes, Scots—anyone in the international business of sea fishing, or seafarers with no particular concern with fishing at all. For all are welcome.

In January, 1970, the R.N.M.D.S.F. in Britain operates twenty hostels, many of them new or extensively rebuilt or modernised within the past ten or fifteen years, and now of a total value in excess of £1,500,000. The annual budget runs at above £250,000, all raised voluntarily, for the organisation is neither state-run nor state-aided. It is a tremendous achievement and enterprise. At a dozen Scots fishing-ports from Oban to Kinlochbervie, from Lochinver to Fraserburgh, Lossiemouth, Peterhead, and all the major fishing-ports of England from North Shields round to Newlyn, the Mission gets quietly and effectively on with its self-appointed task of caring about and caring for the sea-going fisherfolk and their families. The work done afloat was infinite and splendid: now it has long moved ashore where, even more necessary, it has served and grown with the need.

All this has stemmed from the efforts of one man—Ebenezer Mather (whose name was never in *Who's Who*, nor anything of that sort) and his voyage in a Billingsgate fish-carrier to the North Sea smack fleets in 1881. The flag he hoisted has been served and multiplied by many fine men ever since and now flies quietly but proudly wherever there is need, carrying the message that Christ first brought to fishermen almost 2,000 years ago—carrying the message, and serving the ideals.

Chapter 10

FAMILY FISHERMAN

As the *Arctic Vandal* came in the lockpit to the St. Andrew's dock from yet another Arctic voyage and her mariner-fishermen looked to the lines in their best gear, with polished faces and a new look to their eyes, I wondered again about this tremendous and so greatly demanding industry of deep dea fishing carried on now at a rate never before approached, not only from Britain and Europe. The *Vandal* voyage was far from my first contact with it, though I have never been a trawlerman. I'd made a middle waters voyage in the *Samarian* of Grimsby, fishing mainly round the Faeroe Islands, and with a grand crew of wonderful old lads all well past 60 in the drifter *Radium* (built in 1904) out of Stornoway looking for herring in the Minches: I've been with the Portuguese Grand Bankers and Greenland fishermen, the beach fishermen of Albufeira, Palheiros de Mira, Vieiria de Leiria, Nazaré, the tunnymen of Faro in the Algarve. I'd seen the small fish-hunting ships of Japan far and wide in the Pacific north and south, the Russians in the Behring Straits and off the coast of Scotland; my countrymen the

Australians after tunny out of Port Lincoln, S.A. and Eden, N.S.W., the huge fish-hunting fleets from California, Peru, Chile, and the Gulf of Mexico.

What a gigantic, remorseless business it all is—men and ships in their untold thousands, all in the quest for fish and more fish—poor fish!

Scientists and governments fight with such problems as there may be of over-fishing, depletion of stocks, spoilation of grounds. I have been concerned with men—all manner of men from the Brazilian on his *jangada* balsa raft fishing from a beach at Recife—fishing simply, taking little, serving defined local needs, leaving stocks—to my shipmates of the distant-waters trawlers of Hull.

We'd passed huge docks down-river devoted to the trade of overseas steamships. A 6,000-ton car-ferry, shoving half the Humber aside with her hurrying bow, dragging a wake of such disturbance the *Vandal* jumped in it as if stung, had passed us by. Now we were by the fish dock, and moved in. A big stern-freezer bellowed at us in the lock, outward bound and impatient to be gone. Another was swinging with tugs in the river, adjusting compasses. Others of the big boys were berthed alongside, discharging solid blocks of ice and cleaned cod with a rattling apparatus that dragged the iron-hard stuff up from cavernous holds and shot it endlessly into cold storage. One man tended this contrivance. Nobody was about the big fellow's decks, for the crews were ashore to enjoy a leave that never can be long enough.

I wondered again, looking at these big ships, whether they were not an adequate answer to the difficult problems of Arctic fishing safety. Obviously, the big, high-freeboard hulls had built-in advantages over the old-style trawlers: men working in shelter, too, were better protected than those still toiling in open, fish-filled pounds on the low fore-decks. But seamen know that no ship that goes to sea can ever be wholly safe from all hazards. That very morning the radio was full of news about some bloated great new oil-tanker, said to be over 200,000 tons, which in a calm sea off the west coast of Africa had had an explosion aboard which so damaged even that enormous hull that it had sunk. The tanker was bound outwards in ballast round Good Hope for the oil fields of the Arabian Gulf. The great hull was perfectly compartmented. There was no cargo. Men had died. Others were adrift in boats from a laden Liberian tanker which had broken in two somewhere in the farther North Atlantic, and they were never found.

Skilled men do their best to produce good ships but they have won no final victory, nor ever will, with any type.

Across the dock a line of picturesque side-winders like the *Vandal*—those lithe, classic but perhaps now old-fashioned ships of pre-eminent sea-worthiness and vast endurance, their storm-fighting, sea-skill built into them from the savage lessons of countless generations of forbears both ships and men—waited a day or two for time to collect their crews again, and go out to the fight once more. Ice poured down metal chutes into them, fuel-oil pumped into them, maintenance gangs slapped paint on them. Busy hands rushed their rescue-rafts to the depot ashore maintained by the owners' organisation, for thorough overhaul, to be certain that if ever the grim moment came when a toggle grabbed hastily by night meant life if it worked, death if it did not, then that toggle would work and work perfectly, and the life-raft inflate itself ready to float off with four or six men equipped to guide rescuers to them and sustain them while they waited. Later I visited the raft inspection team in their workshops, noted the plaque on the wall which recorded that these rafts had saved 164 lives from trawlers—wrecked, foundered, or stranded—since the owners first introduced them, aware of their worth from war-time and flying experience, well ahead of government action. One hundred and sixty-four lives just from lost trawlers! The names of the ships were all there. Rafts may save when boats cannot. It is a remarkable record.

Other small parties hurried huge bundles of charts ashore from the charthouse, to the owners' co-operative correcting room where a man and a girl, skilfully working, would correct them up to the moment. Welders worked at a sill to raise it half a foot or so in accord with the recommendation of the latest committee of safety to keep the sea more surely out of the quarters. Determined men worked upon generators, main engines, fish-loops, electronic navigators, lights, the trawl-winch, otter-boards, warps and gallows as necessary—all purposeful, efficient, and rapidly. The personnel manager from the office directed the galley-boy, having served the appointed time in that capacity, survived seasickness and reported upon satisfactorily, to attend at the Nautical

College of Kingston-upon-Hull to receive instruction for deck duties: in the meantime the personnel man assured himself that the pale lad was robust and fit. He would be checked thoroughly in due course by the industry's medical department. Fire and accident prevention officers looked each trawler over during her stay in port to see that every precaution against both types of calamity had been taken, and fire-fighting gear was in place and effective.

Meanwhile the *Vandal* had nosed silently into her landing berth to await the morrow's market. In good time, the 'bobbers' would open the fish room, rig their electric winches to whisk the kits of good Arctic fish ashore, where large silent men in white coats and clumping clogs would dispose of them swiftly to the highest bidder, whispering in one ear. I'd watched them at their mysterious calling, disposing of four large trawler loads—over 10,000 kits that morning—within moments. Then the fish were bundled off by refrigerated lorry to the markets, fish-shops, clubs and restaurants of inland England. In the meantime the take of cod-liver oil rendered from fresh livers that voyage is tanked off to the big plant at Marfleet where the bottled product helps to keep babes and adults healthy as far away as Chile, Peru, Alice Springs, and Mahé in the Seychelles Islands. A large wall-map flag-studded for centres of distribution for cod-liver oil and byproducts, shows the amazing variety of cities and countries which have learned to make good use of this worth-while product.

Hull trawler-owners established the plant at Marfleet in

1935: four years later Grimsby joined them in the co-operative enterprise. Today the company produces about one-third of the world's supply of cod-liver oil, and exports a fifth of this. The oil itself has been in use since the 18th century when it helped to treat bone diseases among the poor. In the Second World War, with fresh orange juice and milk, supplies were distributed free for Britain's babies and small children, and played a great part in giving them a healthy and properly nourished start in life.

An astonishing, world-wide business the little *Vandal* serves! In the meantime her people quickly disperse to their homes, many of these not far from the dock though Skipper Bernie lives outside Hull, at Hessle, where owner Tom Boyd also has his home. The Skipper has paper work to do before reunion with Isobel his wife and their four children. The trawlermen, I have no doubt, will re-enter their homes (and do much else in their daily lives, ashore and afloat) with due regard for the appropriate superstitions. They will all, I know, first make certain to go in by the door through which they left, just as when leaving none looked back at home or loved ones, but strode ahead, sea-bag on shoulder, eyes firmly fixed towards the river. Perhaps he'd remembered, too, to remind his wife to do no laundry on the day of his departure, for this can also bring misfortune.

Brian told me of these superstitions and many more. Such things are handed down and observed 'just in case' there might be something in them. Deckie-learners and young hands quickly learned not to leave a broom around the decks while on the grounds in case it swept their luck away. (But this, though useful also for keeping a tidy deck, might date back to the days of belief in trolls and witches, not so long ago. Brian and the others didn't know.) The wives, poor women, pay more attention to the more or less hereditary superstitions even than their men do: they have so much to lose. For them, too, such notions are 'just in case'. Their real faith and support are more surely founded.

It was an owner, I knew, who first suggested to Ebenezer Mather that he might find worth-while work among the North Sea fishermen—Samuel Hewett, a former smacksman himself who, by industry and intelligence, rose from small beginnings to own a large fleet out of the port of Yarmouth, on the coast of Norfolk. Owner Hewett arranged the trip by carrier to the Dogger banks, and throughout his life afterwards was a firm supporter of the mission which followed. So are the owners of the big fishing ports today: so indeed are considerable numbers of the public of Britain. In Hull Tom Boyd is one of the owners who serve on the local committee of the R.N.M.D.S.F. The mission's fine place there is of value to owners, of course, with many new recruits coming to the big fishing port and nowhere to go, and a considerable proportion of out-of-town men in their crews: it is of value in other ways too, particularly as a service both for the men at sea and for their families.

An owner has his hands full in the difficult days of the early 1970's, with the problems of the trade (such as the ever-present threat of underselling from abroad, the changing habits of the public, the enormous and ever-increasing costs of building fishing-ships and catching fish). He is indeed most grateful for the Mission's help as morale builder, welfare agency, and sound and reliable influence for good. The owners of Hull founded their Widows and Orphans' fund back in the 1890's when there were no other such facilities: they still maintain it. They set up a mutual insurance to keep costs down. They put expensive life-rafts and radio-aids in their ships long before the responsible officials thought of insisting on their provision. They endeavour to do the best possible for their men.

Missioner at Hull at the time was Senior Superintendent D. MacMillan. He had plenty to do. A few weeks earlier, his Mission Chapel had been filled three times upon the one day, with the relatives of Hull fishermen recently lost at sea. That day Sir William S. Duthie, O.B.E., indefatigable chairman of the whole Mission, had unveiled at separate services plaques to the memories of the 59 men lost in the Hull trawlers *St. Romanus*, *Kingston Peridot*, and *Ross Cleveland*—a sad occasion indeed, yet the eyes of the bereaved shone also with faith and hope.

Sir William Smith Duthie, O.B.E., son of a fisherman, a Banffshire man (born there in 1892, the year Grenfell joined the R.N.M.D.S.F. which Sir William has headed for many years) has had the wide experience that leads to knowledge of men. Severely wounded with the Gordon Highlanders in the First World War, he saw service in the Canadian Army too, was Deputy Chief for U.N.R.R.A. with the Balkans Mission in '45 after important work ensuring Britain's bread supplies during the Second World War, was M.P. for Banffshire from '45 until '64, when he was very active on behalf of fishermen and fishing interests. A sailor in his own right (he is an experienced yachtsman and an amateur archaeologist too), he knew as a boy the sturdy fishermen of his own N.E. Scotland. The respect and understanding of them he learned then he has never forgotten.

I met him in the modest headquarters of the Mission at 43 Nottingham Place (almost across the road from Madame Tussaud's) in London, a small building where great works are done. I noted the model of a North Sea smack in the hall, the ship's wheel from one of the so useful and hard-working assistance vessels: the plaque with the names of the chairmen dating back to Founder Mather—so few of them, for there has been continuity of purpose and performance here—the efficient young women working at files, computers, the little board room where the members of the Council meet. Admiral Sir Charles Madden, Bart., G.C.B., is Deputy Chairman, Conrad D. Warner, F.C.I.S., treasurer, and members include the Archbishop of York and Lord Wakefield of Kendal.* Senior executives are Messrs. Charles Laurie, secretary, and Jack C. Lewis, O.B.E., J.P.

It is during the period of Sir Wm. Duthie's dynamic chairmanship, I knew, that the great expansion in the Mission's activities has been so successfully undertaken.

---

*Full list of Council members since 1954 may be found in an Appendix, together with a list of the Mission's centres.

A welcome increase in legacies, subscriptions, and donations has gone with this.

"The British public is with us in our efforts," Sir William told me with a smile. "It has great practical sympathy with its fishermen, and it knows that our work is for all, irrespective of creed, colour, or nationality, or anything else. The Lord has honoured our venture."

During 1968 alone, new buildings were opened or old ones modernised at Lowestoft, Lochinver, Newlyn and North Shields, while the splendid new Queen Mary Institute at Grimsby (opened in late '67 with its quiet chapel, excellent amenities, and 51 cabin bedrooms) continued to serve a great need, along with the hostel at Hull and a dozen other centres. At Kinlochbervie, the Institute was being reconstructed. Fraserburgh is next.

Sir William, the Council, the Officers, the Executive, the Missioners in the ports and all the great army of helpers everywhere have much to thank the Lord for, that He has helped them to do so much for the hard-working, hard-living fishermen. There are other organisations also doing good work, too.

The Mission under God does all it may of Christ's work for the fishermen, with the great inarticulate mass of the people of Britain, aware of the fight of these fishermen and the casualties of the battle in their daily lives, behind them too.

There is, I observed, no such man as a typical Missioner. These are trained, dedicated and devoted men for whom life is no bed of roses. They come from the established

Lowestoft fishermen about 1890

(By permission of the Trustees, National Maritime Museum)

Sunday service in the North Sea fishing fleet: from the *Illustrated London News*, November 4, 1882

(By permission of the Trustees, National Maritime Museum)

Church, from the Salvation Army, from fishing, from a great cross-section of life. But they all come because they feel the value of their work, and they know the need.

"The Mission man becomes inescapably involved with his work. He finds himself caught up in the comradeship and brotherhood which is unique among fishermen," said Senior Superintendent H. L. Moore of Grimsby, a man who came from the church because he saw the need. "Over a period of time the fishermen become involuntarily a part of us, just through our constant contact. This is their Mission—to some their home too, whether they come in for a cup of tea, a game of darts or billiards, or just to sober up . . .

"The fishermen and their families have a part of us. We see the ice floes in the harbour and the river, hear the matter-of-fact voice over the radio announce winds up to 100 miles an hour . . . We have a good idea who is out there . . ."

Often, too often, the man in the Mission uniform must bring terrible news to little homes, of ships overdue, then lost. To home after home the news must go. Better to be told by a sympathetic missioner than to see first tidings perhaps garbled in a newspaper or mentioned casually on the air.

I talked with Superintendent Jim Ralph, aged 29, former Lossiemouth fisherman now in charge of the new Mission at Lerwick in the Shetland Islands. From the front windows of the fine building on the hill, we looked down on the harbour where the sturdy seine-netters and long-liners of Wick and Scallaway, Scrabster, and Peterhead jostled with sturdy little Norsemen and Danes, secured in serried rows beside stone quays. Among them was the shapely little *Skömvaer II*, a Norwegian assistance vessel based on Lerwick. A big Russian with a semi-icebreaker bow and a long, sturdy hull—a fresh-water tanker for an outside fleet which never came in—lay at another berth, taking in drinking water for perhaps a hundred ships, or some factory-ship too large for the little port. Snow covered all the tree-less hills and dusted the narrow streets where sea-booted, fur-capped fishermen stretched their legs, regardless of the storm.

What had caused him to leave fishing two years earlier, at the age of 27, I asked Jim, sitting there very quietly in the pleasant room.

"I don't think I really have left fishing, here," he replied, at first. Nor, indeed, had he.

James Ralph was just as surely born into fishing at Lossiemouth in 1940 as any lad into a trawlerman's family at Hull, or Grimsby, or Lowestoft, or Aberdeen. As is the way of things in north-east Scotland, by dint of hard work and his own long sustained efforts, his father—himself the son and grandson of fishermen—had acquired a sea-going fishing vessel. He named her *Devotion*, and off to sea Jim went in her at the age of fourteen. She was a family vessel, like so many more of the Scots—a vessel the menfolk of the family had worked and saved for over half a lifetime, and used for the family living. Fathers and sons worked such ships, though they knew quite well that the sea could and one day might take them all. If there were not enough sons—Jim was an only son—cousins and uncles joined.

"I know it is said—especially after losses—that families should be stopped going to sea in the one ship," said Jim. "But I don't think that can ever be stopped in Scotland. Ships cost so much. We must work hard to clear costs and pay our way. But we are working for ourselves. You can see where you are going. Some of the stress of modern life and fishing is avoided that way."

The *Devotion* fished as far away as the Shetlands and St. Kilda, summer and winter, 50 weeks a year. She did well.

Then one day in 1960 the little ship was wrecked—flung up on rocks below a cliff in a hard blow with bad visibility. The seas smashed her in an instant, and the rocks ground her to pieces. And drowned half the crew including his father, aged forty-nine.

Jim was not sure how he survived that bitter night. Flung out on mussel-covered rocks which tore his hands and arms and lacerated his feet, wet through, chilled, almost exhausted, he stumbled for miles. At last he found a place where he could climb the cliffs.

He went right back to sea. He was 19, the man of the family now—the only man. The new crew were more uncles and cousins, and the little *Ocean Gleaner* was a hard-working, happy ship, as the *Devotion* had been.

One day in the Minches, a sea suddenly leapt up and washed aboard, in the way seas have. When it had gone, so had a cousin.

"I was talking with him one moment. Next, he was

in the sea. We almost got to him to haul him aboard. But he sank, and he did not rise again. We saw him no more."

Aged 19: one of so many!

Jim knew about the Mission. Every fisherman did. He was a Christian. He decided that, perhaps, here was work that he could do. There was. He began as an assistant at Hull. One of his first jobs there was to be sent round with Mr. MacMillan to break the news to the families of the men missing in the *St. Romanus*, *Ross Cleveland*, and *Kingston Peridot*.

At one house was a mother who had already lost two fishermen sons. She was sitting alone, staring into a bit of a fire. She knew why they had come. Now the third, and last . . .

She gave no sign that she heard anything that was said. She just sat there, staring into the fire.

A SHORT BIBLIOGRAPHY

Books of use in the preparation of this work included the following:

Nor'ard of the Dogger, by E. J. Mather. London, John Nisbet & Co., 20 Berners Street. 1887.

The Sea-Fishing Industry of England and Wales, by F. G. Aflalo, F.R.G.S., F.Z.S. London, Edward Stanford, 12, 13, 14 Long Acre, W.C.2. 1904.

North Sea Fishers and Fighters, by Walter Wood. London, Kegan Paul, Trench, Trubner and Co. Ltd. 1911.

Fisheries of the North Sea, by Neal Green. London, Methuen & Co. Ltd., 36 Essex St., W.C. 1918.

Wilfred Grenfell, His Life and Work, by J. Lennox Kerr. London, George G. Harrap and Co. 182 High Holborn, W.C.1. 1959.

Sailing Trawlers, by Edgar J. March. Newton Abbot, Devon: David and Charles. Reprint, 1969.

Conditions of Work in the Fishing Industry, International Labour Office, Geneva, 1952.

Posted Missing, by Alan Villiers. London, Hodder and Stoughton Ltd., 1956 and (Paperback) 1959: also Charles Scribner's Sons, New York, U.S.A.

Quest of the Schooner Argus, Alan Villiers. London, Hodder and Stoughton Ltd. and Charles Scribner's Sons, New York, U.S.A.

Introduction to Trawling, by A. Hodson (A text book). Richardsons and Copping Ltd., Grimsby 1948: Revised edition reprinted 1967.

Modern Deep-Sea Trawling Gear, by John Garner. London, Fishing News (Books) Ltd., 110 Fleet Street, E.C.4. 1967.

Fish Catching Methods of the World, by Dr. A. von Brandt. London, Fishing News (Books) Ltd., 110 Fleet Street, E.C.4. 1964.

Trawler Safety: Final Report of the Committee of Inquiry. Chairman, Admiral Sir Deric Holland-Martin. London, H.M. Stationery Office 1969. (Cmd. 4114).

Toilers of the Deep, Quarterly Journal of the Royal National Mission to Deep Sea Fishermen. 43 Nottingham Place, London, W1M 4BX

APPENDICES

APPENDIX I

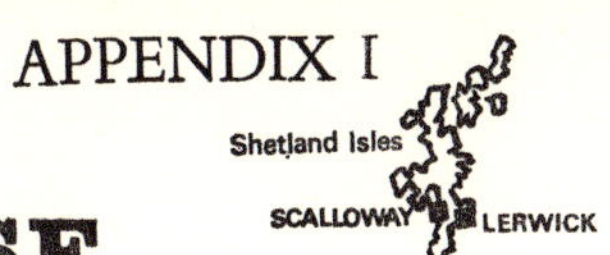

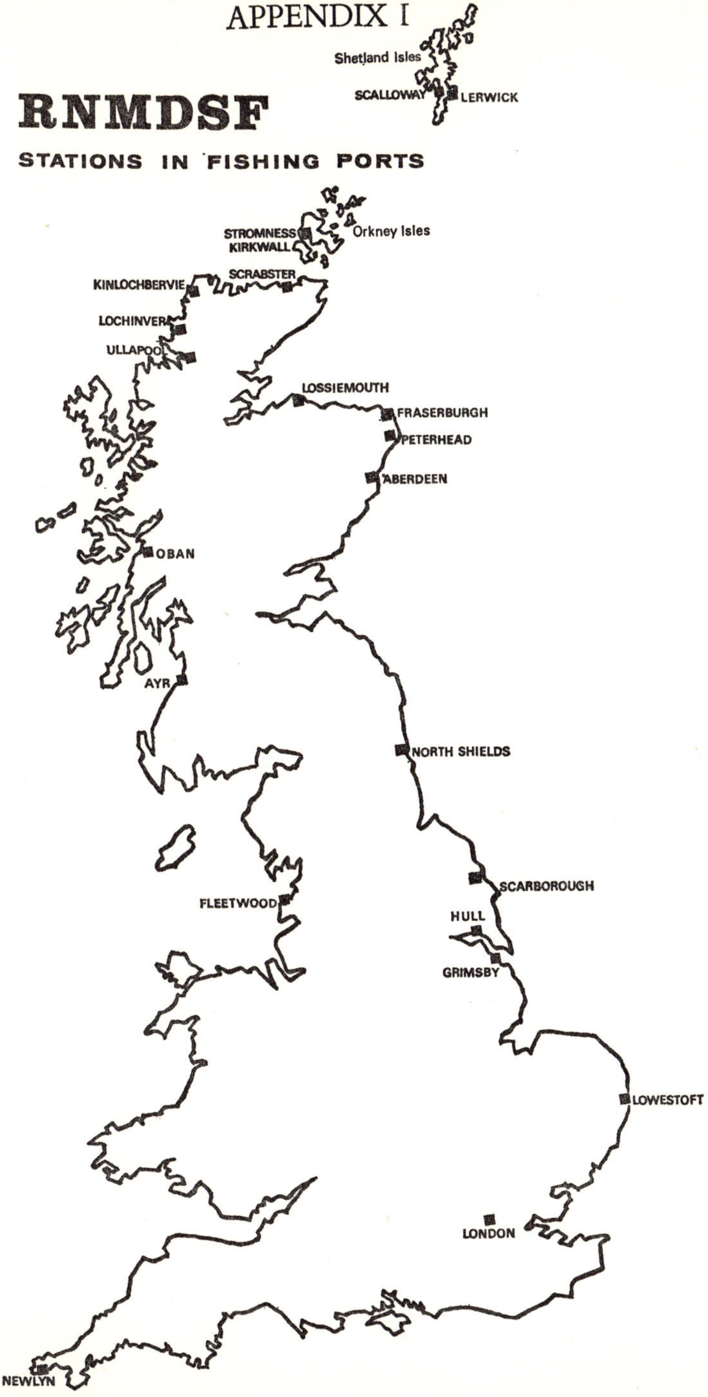
RNMDSF
STATIONS IN FISHING PORTS
Shetland Isles
SCALLOWAY
LERWICK
STROMNESS
KIRKWALL
Orkney Isles
SCRABSTER
KINLOCHBERVIE
LOCHINVER
ULLAPOOL
LOSSIEMOUTH
FRASERBURGH
PETERHEAD
ABERDEEN
OBAN
AYR
NORTH SHIELDS
SCARBOROUGH
HULL
GRIMSBY
FLEETWOOD
LOWESTOFT
LONDON
NEWLYN

APPENDIX II

Royal National Mission to Deep Sea Fishermen centres at March, 1970, are at the following places:

R.N.M.D.S.F.,
Palmerston Road,
ABERDEEN.

R.N.M.D.S.F.,
South Harbour Street,
AYR.

R.N.M.D.S.F.,
Dock Street,
FLEETWOOD.

R.N.M.D.S.F.,
73 Broad Street,
FRASERBURGH,
Aberdeenshire.

R.N.M.D.S.F.,
Hope Street,
GRIMSBY,
Lincs.

R.N.M.D.S.F.,
Goulton Street
HULL,
Yorks.

R.N.M.D.S.F.,
Fish Pier,
KINLOCHBERVIE,
Sutherland.

R.N.M.D.S.F.,
Fort Charlotte,
LERWICK,
Shetland Isles.

R.N.M.D.S.F.,
Culag Park,
LOCHINVER,
Sutherland.

R.N.M.D.S.F.,
Commerce Street,
LOSSIEMOUTH,
Morayshire.

R.N.M.D.S.F.,
Waveney Road,
LOWESTOFT,
Suffolk.

R.N.M.D.S.F.,
NEWLYN,
Penzance,
Cornwall.

R.N.M.D.S.F.,
Union Quay,
NORTH SHIELDS,
Northumberland.

R.N.M.D.S.F.,
Railway Pier,
OBAN,
Argyllshire.

R.N.M.D.S.F.,
Union Street,
PETERHEAD,
Aberdeenshire.

R.N.M.D.S.F.,
Blackness Pier,
SCALLOWAY,
Shetland Isles.

R.N.M.D.S.F.,
The Bethel,
Sandside,
SCARBOROUGH,
York.

R.N.M.D.S.F.,
West Quay,
SCRABSTER,
Caithness.

R.N.M.D.S.F.,
Quayside,
STROMNESS,
Orkney Isles.

R.N.M.D.S.F.,
Shore Street,
ULLAPOOL,
Ross-shire.

APPENDIX III

Council Members of the Royal National Mission to Deep Sea Fishermen and their terms of office from 1954:

Sir William S. Duthie, O.B.E. . .	1954–
Vice-Admiral Sir W. G. Agnew, K.C.V.O., C.B., D.S.O. . . .	1954–1960
Lord Wakefield of Kendal, M.A. . .	1954–
Rev. Rowland Hill	1954–1968
Rev. I. R. N. Miller, M.A. . . .	1954–
Captain J. S. Clarke, V.R.D., M.A., R.N.R.	1955–
Air Marshal S. C. Strafford . . .	1959–1966
Mr. C. D. Warner, F.C.I.S. . .	1960–1969
Sir J. Croft Baker, C.B.E. . . .	1961–1962
Mr. H. Atkinson, O.B.E. . . .	1963–
Mr. P. J. Ensor	1965–
Admiral Sir Charles Madden, Bt., G.C.B., D.L.	1966–
John R. Noble	1966–
The Rt. Hon. The Lord Archbishop of York, Dr. F. D. Coggan . . .	1969–
Mr. Basil Parkes, O.B.E., J.P. .	1969–

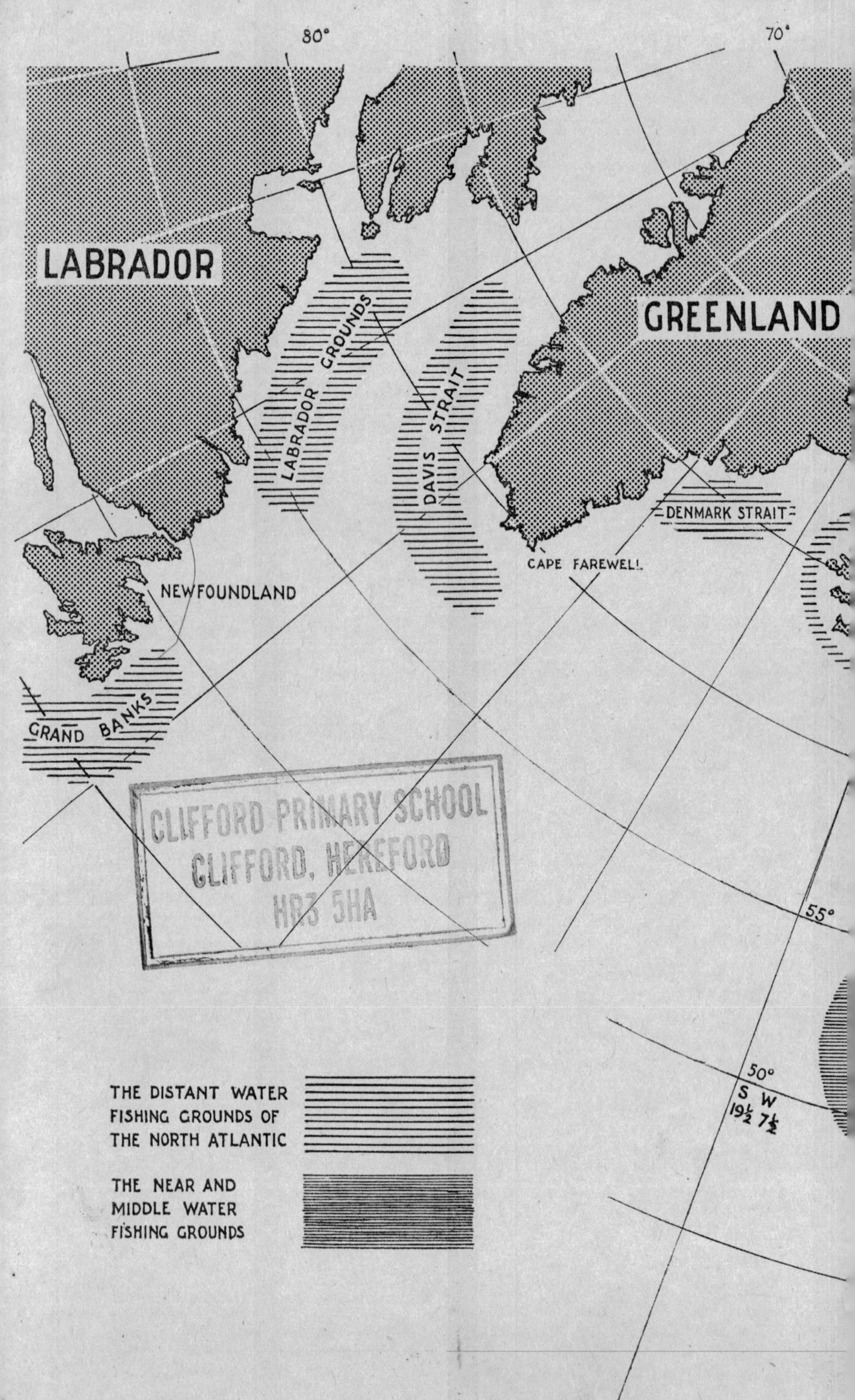
80°
70°
LABRADOR
GREENLAND
LABRADOR GROUNDS
DAVIS STRAIT
DENMARK STRAIT
CAPE FAREWELL
NEWFOUNDLAND
GRAND BANKS
55°
50°
S W
19½ 7½
THE DISTANT WATER
FISHING GROUNDS OF
THE NORTH ATLANTIC
THE NEAR AND
MIDDLE WATER
FISHING GROUNDS